Public–Private Partnerships in International Construction

Over the last ten years, public–private partnerships (PPPs) have become ever more popular worldwide, expanding the body of experience among construction professionals, government agencies and industry. In these economically challenging times, public–private partnership (PPP) has emerged as a crucial framework for providing infrastructure, and also to boost construction industry activity, while shielding the taxpayer from some of the cost. Understanding the lessons learnt is essential to ensuring the success of future projects, and this timely book will prepare the reader to do just that.

Starting by defining PPP itself, Part I is designed to help the novice to get to grips with the basics of this topic. Part II tackles the practicalities of PPPs, including successful implementation, managing the risks involved and how to assess the suitability of a project for the PPP route. Part III presents detailed case studies from Asia, Africa and Australia to illustrate how PPPs should be managed, how problems emerge, and how PPPs can differ across the world.

Drawing on extensive internationally conducted research, from both industry and academia, the authors have written the essential PPP guide. Taking into consideration the perspectives of those in the public sector and the private sector, as well as built environment professionals, it is essential reading for anyone preparing to work on public–private partnerships in construction.

Prof. Albert P.C. Chan, Associate Dean, Faculty of Construction and Environment; Professor, Department of Building and Real Estate, the Hong Kong Polytechnic University, Hong Kong

Dr Esther Cheung, Programme Manager, Housing and Built Environment, College of Humanities and Law, School of Professional and Continuing Education, the University of Hong Kong, Hong Kong

Public–Private Partnerships in International Construction

Learning from case studies

Albert P.C. Chan and Esther Cheung

Routledge
Taylor & Francis Group

LONDON AND NEW YORK

First published 2014
by Routledge
2 Park Square, Milton Park, Abingdon, Oxon, OX14 4RN

Simultaneously published in the USA and Canada
by Routledge
711 Third Avenue, New York, NY 10017

Routledge is an imprint of the Taylor & Francis Group, an informa business

© 2014 Albert P.C. Chan and Esther Cheung

The right of Albert P.C. Chan and Esther Cheung to be identified as author of this work has been asserted by them in accordance with sections 77 and 78 of the Copyright, Designs and Patents Act 1988.

British Library Cataloguing in Publication Data
A catalogue record for this book is available from the British Library

Library of Congress Cataloging-in-Publication Data
Chan, Albert P. C.
Public private partnerships in construction : learning from case studies / Albert P.C. Chan and Esther Cheung.
    pages cm
    Includes bibliographical references and index.
    ISBN 978-0-415-52975-4 (hardback : alk. paper) -- ISBN 978-0-203-73666-1 (ebook) 1. Public-private sector cooperation. 2. Public-private sector cooperation--Case studies. 3. Construction projects--Finance. 4. Public works--Finance I. Cheung, Esther. II. Title.
    HD3871.C47 2014   624.068--dc23
    2013006577

ISBN13: 978-0-415-52975-4 (hbk)
ISBN13: 978-0-203-73666-1 (ebk)

Typeset in Goudy by
Bookcraft Limited, Stroud, Gloucestershire

Printed and bound in Great Britain by
TJ International Ltd, Padstow, Cornwall

Contents

PART III
Public–private partnership case studies   **123**

Figures

Tables

Preface

Many governments have been suffering from significant budget deficit since the last global financial turmoil. Under such a tight budget shadow, many have taken the initiative in radically increasing the private sector involvement in the delivery of public services and infrastructure to the community. Public–private partnerships (PPPs) are collaborations where the public and private sectors both bring their complementary skills to a project, with different levels of involvement and responsibility, for the sake of providing public services.

Over the years public–private partnership (PPP) application has extended from the traditional transportation type projects to more complex social projects such as art and culture facilities. The financial arrangements have also been repackaged into different forms and types, where more possibilities have been revealed. Traditional PPP projects have very much relied on full private financial support whereas modern projects have shown that PPP projects can also be supported partially or solely by the public sector.

PPP has been practised in many developed countries in Europe, North America and Australasia for delivering construction and building projects. But unfortunately not all of these PPP projects have been equally successful. For countries that are new at adopting PPP it is important for them to identify the critical success factors in order to maximise the advantages of this method and to allocate the risks of the concerned parties equitably.

Risk is inherent and difficult to deal with in PPP projects and requires a proper risk management framework. Governments procuring a PPP project would specify its preference as to how the project risks should be shared; private investors would assess their capability of taking these risks, and then propose a bidding price. The contract negotiation would probably focus on the risk-sharing mechanism. A generally accepted principle is that risk should be allocated to the party best able to manage it and at the least cost.

Therefore this book aims to evaluate the merits and shortcomings of PPP, determine the best condition for adopting PPP, identify a series of critical success factors for implementing PPP based on the lessons learned, as well as to develop an equitable risk allocation scheme for delivering PPP projects.

This book is divided into three parts. Part I of the book consists of three chapters which look at the principles of PPP. The first of these chapters looks at the fundamentals and specific features related to PPP. The second chapter presents six financial models for conducting public works projects where varying levels of public and private sector involvement can be seen. Chapter 3 examines the development of PPP on an international level by looking at the past, present and future of PPP across five different continents. Part II of the book consists of five chapters looking at the perspective of PPP according to different parties. Chapter 4 looks at the views from the public sector, private sector and researchers; a comparison between different countries was also conducted. Chapter 5 presents the findings of a questionnaire survey conducted with practitioners which identifies the attractive and negative factors of PPP. Chapter 6 continues to present other findings from the same questionnaire, including the reasons for implementing PPP projects, the factors for successful PPP projects, and also the measures to enhance value for money in PPP projects. Chapter 7 illustrates an evaluation model for assessing the suitability of PPP projects using a high-profile case study. Chapter 8 identifies and ranks the risk factors associated with using PPP. Part III of the book presents three chapters which analyse some interesting PPP case studies from around the world. Chapter 9 looks at an innovative type of PPP where projects are financed solely by the public sector, and where the private sector is involved for other benefits. Chapter 10 looks at some less successful case studies where valuable lessons can be learnt. The final chapter, Chapter 11, looks at the struggles of using PPP in the developing world and how different variations of the model have been attempted.

We would like to take this opportunity to thank those who have supported us throughout the course of this book and contributed towards its completion. Without them this book would definitely not have been possible. These include Dr Daniel W. M. Chan, Dr Edmond W. M. Lam, Dr Kim-wah Chung, Dr Patrick T. I. Lam, Miss Zoe Wang, Mr Ernest E. Ameyaw, Mr Tong Peng, Dr Ye-lin Xu, Prof. Yat-hung Chiang (Hong Kong Polytechnic University, Hong Kong), Dr Bing Li (Xiamen University, China), Dr Chi-pang Lau (Lingnan University, Hong Kong), Dr John F. Y. Yeung (Hong Kong Baptist University), Dr Yong-jian Ke (The University of Newcastle, Australia), Prof. Akintola Akintoye (University of Central Lancashire, UK), Prof. Bo Tang (University of Hong Kong, Hong Kong) and Prof. Shou-qing Wang (Tsinghua University, China).

The financial support from the Research Grants Council of the Hong Kong Special Administrative Region in funding the following research projects upon which the content of this book is based is also gratefully acknowledged:

Developing a Best Practice Framework for Implementing Public Private Partnerships (PPP) in Hong Kong (RGC Competitive Bids) 2005–08.

Developing an Equitable Risk Sharing Mechanism for Public-Private Partnership (PPP) Projects in the People's Republic of China (PRC) (RGC/NSFC Joint Research Grant) 2008–10.

Evaluating the Social, Economical, Cultural and Heritage Impacts of the 'Revitalising Historic Buildings through Partnership Scheme' in Hong Kong (RGC Public Policy Research) 2010–11.

Evaluating the economic, social and cultural impacts of revitalizing industrial buildings in Hong Kong (RGC Public Policy Research) 2012–14.

Finally, this book would not have been possible without sponsorship from the Hong Kong Institute of Surveyors.

Foreword

Public–private partnerships have continued to play an important role in the construction industry. The success of public–private partnerships has been a result of combining the best of the government and the private sector to provide better public projects. Public–private partnerships have also demonstrated huge risks and failures. Consequently, countries are both keen and concerned about their adoption.

This book is a great read for academics and practitioners as well as students. It is based on the fundamental concepts of public–private partnerships which are illustrated by both successful and unsuccessful real life case studies from around the world. The book is divided into three main parts. The first covers some general background, principles and history of public–private partnerships. The second part looks primarily at what different stakeholders say about the approach such as the differences in views between stakeholders, the consolidated reasons for implementation, the success factors, the approaches for enhancing value for money, evaluation of projects and the potential risks involved. The last part of this book presents case studies from Asia, Australia and Africa, including innovative examples, poorly conducted projects and also the approach in the developing world.

It is my pleasure to highly recommend this book to those that are both experienced and inexperienced with public–private partnership projects. The book is easy to grasp for beginners but also resourceful and enlightening for more experienced readers.

Sr Stephen Lai
President, The Hong Kong Institute of Surveyors
February 2013

Abbreviations

ACRHB	Advisory Committee on Revitalisation of Historic Buildings
AVRL	Aqua Vitens Rand Ltd
BOO	Build Own Operate
BOOR	Build Own Operate Remove
BOOT	Build Own Operate Transfer
BOT	Build Operate Transfer
BTO	Build Transfer Operate
CCT	Cross City Tunnel
CHT	Cross Harbour Tunnel
DBFO	Design Build Finance Operate
DBFOM	Design Build Finance Operate Manage
DBO	Design Build Operate
df	degree of freedom
GWCL	Ghana Water Company Limited
GWSC	Ghana Water and Sewerage Corporation
HK$	Hong Kong dollars
HKYHA	Hong Kong Youth Hostels Association
HKZMB	Hong Kong–Zhuhai–Macau Bridge
JV	joint venture
KPI	key performance indicator
l/hd	litres per head
LROT	Lease Renovate Operate Transfer
NDRC	National Development and Reform Commission
NGO	non-governmental organisation
O&M	Operation and Maintenance
OM&M	Operate Maintain and Manage
PFI	Private Finance Initiative
PPP	public–private partnership
PPPs	public–private partnerships
PURC	Public Utilities Regulatory Commission
RMB	renminbi
S	score assigned by user for individual factor within factor group
SCAD HK	Savannah College of Art and Design, Hong Kong
SETP	South Bank Education and Training Precinct

TAFE	technical and further education
TS	total score of factor group
US$	American dollars
VFM	value for money
W	Kendall's coefficient of concordance
WATSAN	Water and Sanitation Committees
WHC	Western Harbour Crossing
WKCD	West Kowloon Cultural District
WSDB	Water and Sanitation Development Board
Wt	weighting of individual factor within the factor group

Part I
Public–private partnership principles

1 Fundamentals and features of public–private partnership

Introduction

Public–private partnership (PPP) is a procurement approach where the public and private sector join forces to deliver a public service or facility. In this arrangement normally both the public and private sector will contribute their expertise and resources to the project and share the risks involved. The definition of PPP may differ slightly among different jurisdictions, depending on which part of the arrangement the importance is focused on. For example, PPP is defined as any agreement where the public and private sectors work together to deliver a public project: 'Arrangements where the public and private sectors both bring their complementary skills to a project, with varying levels of involvement and responsibility, for the purpose of providing public services or projects' (Efficiency Unit 2012a).

Another source describes the term PPP as: 'An arrangement for the provision of assets or services, often in combination and usually for a substantial or complex "package", in which both private sector supplier and public sector client share the significant risks in provision and/or operation' (Infrastructure Implementation Group 2005). In this definition there is an emphasis that both the public and private parties share a large proportion of the risks in a PPP project. In reality it is not always that an equal split of risks is experienced. Naturally, each party will want to pass on more risks to the other party. It is noticed that this occurrence is more common in developing countries or jurisdictions where the government has less experience in this alternative procurement method. Previous publications have indicated the importance of the financing of PPP projects and how passing on financial risks to the private sector is appealing to governments: 'Privately financed projects involve provision by investors of equity capital and debt capital to fund what might otherwise be wholly publicly funded projects financed from Government borrowings and/or budget revenue' (Infrastructure Implementation Group 2005).

This chapter looks specifically at the fundamentals and features of PPP projects in general. The areas considered include a comparison with the traditional practice, some background information, and the attractive factors, negative factors, value for money and critical success factors of PPP.

Traditional versus PPP

The procurement processes for the traditional approach and PPP approach are similar. PPP is a contractual agreement involving the private sector in the delivery of public services. Irrespective of whether the project is adopting a traditional or PPP approach, the procuring government departments should follow the same relevant regulations and procedures.

According to the Efficiency Unit (2008), the steps involved in the procurement of public works projects are as follows:

- Step 1: The client department will define a facility or service that is required. The relevant works department will produce a design. In a PPP approach the design may be very preliminary so that there is room for private sector innovation. In the traditional approach the design would be a lot more concrete.
- Step 2: The client department establishes a Project Steering Committee comprising civil servants, and possibly external experts if required, to monitor the project.
- Step 3: The works department will conduct public consultations and obtain financial endorsement.
- Step 4: After the planning and approvals are obtained, an output-based service specification would be prepared where the private sector will be involved with the delivery of public services.
- Step 5: The project may adopt a two-envelope tender evaluation approach, where the successful bidders should pass all the mandatory requirements and obtain the highest combined scores in the technical and non-technical assessment in general.
- Step 6: For traditional projects, the works department would monitor the construction process whereas in PPP projects the client department deals with the consortium only and the contractors are monitored by the consortium.
- Step 7: When the project is completed, the works department would inspect the works for traditional projects, but for PPP projects the client department or a third party would verify the facility to be fit for purpose. Payment would be made to the private party either directly by the government or by the end-users of the project.

Background of PPP

The evolution

PPP projects can be dated as far back as the 1800s during the railway construction boom in the UK (Grimsey and Lewis 2004). PPP is a relatively modern term for this arrangement used only more commonly in the last decade. Previously, variations of the arrangement included Private Finance Initiative (PFI), which is a more familiar term to many people due to its popular development in the UK during the early nineties (Tieman 2003).

It would not be incorrect to say that the PFI practice developed in the UK raised the world's attention to this alternative option for delivering public infrastructure and services. PPP projects now account for about 15 and 8 per cent of infrastructure spent in the UK and Australia respectively (Ernst and Young 2005). As of 2011, approximately 700 PFI contracts had been secured in the UK, with 500 of these being in England. The combined capital value of these projects is almost £50 billion (HM Treasury 2012). Furthermore, the local Treasury estimates that there will be approximately £200 billion worth of PFI contracts within the next 25 years in the UK. By using the PFI approach, the local Treasury claims that annual savings are around £2–3 billion per year (National Audit Office 2011). However, Maltby (2003) asserted that PPP/PFI should be abolished for smaller projects and for information technology schemes.

Partnership UK was set up in 2000 to succeed the Treasury Taskforce. The Taskforce was set up in 1997 to oversee the implementation of PPP/PFI projects (Partnerships UK 2012a). One observation is that Partnerships UK was initiated by the local Treasury. The team is generally responsible for providing project advice and support, developing government policies, providing co-sponsorship and investing in PPP/PFI projects.

Due to the long history of PPP/PFI projects in the UK, Partnerships UK has a very comprehensive collection of guidelines and policies on implementing PPP projects for all sectors in many aspects. Case study reports can also be found in the public domain. Amongst the projects conducted by Partnerships UK the majority included projects for schools, hospitals and transportation. Other projects which have also been conducted include environment ones, leisure facilities, prisons and detention centres, housing, and so on (Partnerships UK 2012b). The extent to which PFI could be used and the advantages created were the main drivers attracting other countries to start adopting or improve their practice in PPP.

A more specific term used more commonly decades ago is Build Operate and Transfer (BOT). This arrangement was commonly adopted for transportation projects. This is because transportation projects tend to be larger in size and also because their long physical lives fit well into the procurement model. Early types of public infrastructure projects that involved the private sector include the turnpikes built in the UK and the US, and also the water facilities that the French delivered through the concession approach (Grimsey and Lewis 2004). Although water projects tend not to be particularly large in project sum, the advantages were noticed early on of introducing private expertise to deal with tasks that the public sector was probably not as efficient or experienced in carrying out. On the other hand, PPP also plays a significant role in the infrastructure development of developing countries. Figure 1.1 presents the annual private investment between 1990 and 2006 in the public services of developing countries (World Bank 2008).

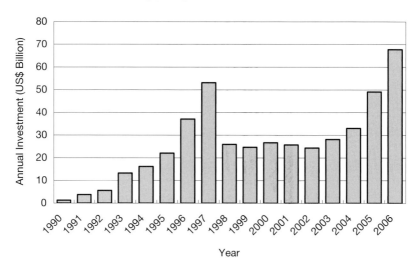

Figure 1.1 Annual investment of infrastructure projects with private participation in developing countries 1990–2006 (World Bank 2012; Cheung *et al.* 2010b) (with permission from Emerald Group Publishing Ltd and the World Bank Group)

Types of PPP

There are many types of PPP used around the world. Most of them operate in similar ways and the name differs depending on the country it is used in, whereas in some cases there are major differences to the approach. Some of the commonly mentioned different types of PPP are now described.

Design Build Finance Operate (DBFO) is similar to BTO; the government will retain title of the land and lease it to the private consortium over the life of the concessionary agreement (Levy 1996).

Operation and Maintenance (O&M) is where the private company operates and maintains a publicly owned asset. This is especially common in mainland China as traditionally the majority of assets are state owned. The large number of state-owned facilities and services have meant that the Chinese government has held a heavy burden, and by adopting PPP this financial commitment can be released. This type of PPP is sometimes not as favourable as ones that start from scratch. For new projects the benefits of employment are obvious, but on the other hand for existing facilities and services a consortium taking over can affect the existing employees.

Private Finance Initiative (PFI) is commonly used in the UK; there is great emphasis on private financing.

Build Operate Transfer (BOT) is one of the most traditional types of PPP used in the early days mainly for transport economic infrastructure projects. This has also been the traditional option used in Hong Kong. BOT involves the construction of the facility as well as its operation. At the end of the contract period it will be transferred back into the hands of the government.

Build Own Operate (BOO) was commonly used in Australia at the start.

Build Own Operate Transfer (BOOT) was also commonly used in Australia at the beginning. It is similar to BOT but with a larger emphasis on ownership.

Build Transfer Operate (BTO) is a method of relieving the consortium of furnishing the high-cost insurance required by the project during operation of the facility (Levy 1996).

Joint venture (JV) describes situations where the public and private sectors jointly finance, own and operate the facility (Grimsey and Lewis 2004).

Leasing is where all or a substantial part of all risks associated with funding, developing and operating the facility are assumed by the private sector, with the public sector entity taking the facility on lease (Sapte 1997).

The PPP process

This section of the chapter looks at the PPP process in detail. In a typical PPP project the government will invite private consortia to bid by submitting a project proposal. The successful bidder will need to design, construct and manage the facility (or service) for the agreed concessionary period, which is typically ten to thirty years. Over the concessionary period the private consortium will need to maintain and operate the facility according to the contract terms defined by the government. Normally certain quality standards or performance targets must be achieved. Part of the profit made from the project will be used to repay the loan that the consortium took out to cover the design and construction costs. The remaining proportion becomes their profit, so obviously it is to their benefit to manage the project well. At the end of the concessionary period the private consortium will normally hand back the facility into the hands of the government.

In general, the typical processes for delivering PPP projects in New South Wales include five major steps: (1) Project identification; (2) Project approval; (3) Planning assessment; (4) Project delivery; and (5) Project implementation (Infrastructure Implementation Group 2005). Before a project is even considered for the PPP path it will go through a series of governmental in-house procedures to decide whether it is a public facility or service that is needed. If deemed to be necessary, the project will have to be approved via the Gateway review process and to see which procurement option it should adopt. Planning assessment via a number of different line agencies would be necessary. Finally the project will be offered to the market, consortia will bid for it and the government will select the most suitable candidate after a long series of negotiations. The project will be designed and constructed typically over three to five years. It will then be operated and maintained for a further twenty-five to thirty years as the concession period. Thereafter, the project will normally be returned to the government, completely ending its life as a PPP project.

The parties involved in a PPP project

In a PPP project there are usually four key parties involved: the local government department (public sector), the consortium (private sector), the employees of the project and also the public and end-users of the facility or service (Figure 1.2).

In traditional PPP projects the government is usually more concerned with transferring the risks associated with design, construction, management, operation and so on to the private sector and satisfying the needs of the general public. On the other hand, the consortium is usually willing to accept some risk but in return expects a more satisfactory financial profit. In the traditional practice the government and the consortium are more self-centred, focusing on their own benefits rather than trying to achieve a win–win scenario. Recent years have shown a change to this practice; the parties are more willing to share responsibility, communication is increased and the partnership apart from being based on finance is also concerned with maximising the benefits that can be adopted from the private sector and bringing in skills and innovations that the public sector does not possess.

The employees of a PPP project benefit through employment. For this group of people a successful PPP project often indicates job security. The general public end-users have been known to have a large effect on the success of a PPP project. Often it is not whether a PPP project is finished ahead of time or is making a huge profit that determines its success. In many cases it is often its image perceived by the general public from the media that is its key to success. Public opinion is important; hence a successful PPP project must consider its overall image. For example, a project that needs to cut down forests for construction may be seen to fail at the beginning due to its lack of environmental awareness. Therefore a strategic plan must be considered at the start, possibly even before the drawing board.

Research conducted in PPP

With the increasing popularity of adopting PPP projects around the world, research in this field has also become more important to both researchers and practitioners (Al-Sharif and Kaka 2004). A comprehensive literature review of PPP research was

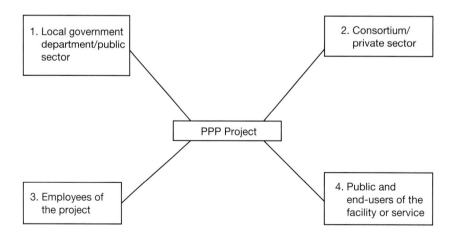

Figure 1.2 Parties involved in a PPP project

previously conducted by Ke *et al.* (2009). A total of 148 recent publications from renowned journals were studied. The findings showed that researchers from the UK were the originators of most PPP papers, followed by the US, Singapore, Hong Kong, China, Australia and Germany. It was assumed that construction education, national economics and mother language were all factors affecting which countries published more PPP papers.

In academic institutions, Nanyang Technological University in Singapore, the University of Hong Kong, National University of Singapore and Glasgow Caledonian University were all identified as active in pursuing PPP research. It was also found that various modes of PPP have been applied in different parts of the world, and the diverse concept of PPP has been publicly accepted instead of the more traditional BOT scheme alone.

PPP topics that were found to be of particular interest to the researchers included 'Risk', 'Procurement' and 'Finance'. Seven more specific categories were derived from these topics including (a) Investment environment; (b) Procurement; (c) Economic viability; (d) Financial package; (e) Risk management; (f) Governance issue; and (g) Integration research. For these research studies, the techniques adopted vary from qualitative to quantitative analyses, some of which have included more rigorous techniques and/or theories in researching.

Attractive factors of PPP

The attractive factors of PPP have been discussed by many previous researchers. This section looks briefly at some of these. So why are governments across the world favouring the approach of PPP to provide for their public services and facilities? The very first PPP projects that opted for this approach were simply to bring in private investment for public services and facilities. These services and facilities were often essential for the public but to provide for them using the government's capital would put pressure on the government's financial status. Therefore, it was an ideal situation that the public had what they wanted provided for without the government having to pay, and also business opportunities were widened for the private sector.

As PPP has developed over the years the associated advantages have become more obvious. Walker and Smith (1995) suggested three main reasons for using the PPP approach:

- In general, the private sector possesses better mobility than the public sector. For example, the private sector is not only able to save the costs of project planning, design, construction and operation, but also avoid the bureaucracy and to relieve the administrative burden;
- The private sector can provide better service to the public sector and establish a good partnership so that a balanced risk–return structure can be maintained; and
- The government lacks the ability to raise massive funds for the large-scale infrastructure projects, but private participation can mitigate the government's financial burden.

In addition, Walker and Smith (1995) maintained that PPP is a win–win solution and recognised a number of benefits to the general public and government:

- Relief of financial burden;
- Relief of administrative burden;
- Reduction in size of (inefficient) bureaucracy;
- Better services to the public;
- Encouragement of growth; and
- Government can better focus and fund social issues such as health, education, pensions and arts.

It is anticipated that there will be more PPP projects, for two main reasons according to Ghobadian *et al.* (2004). First, the private sector will get to know the needs of the public sector client over time. Second, the private sector has more to give than the public sector in terms of skills, technology and knowledge, therefore providing better quality facilities.

Askar and Gab-Allah (2002) summarised eight advantages of PPP in their paper:

- The use of private sector financing to provide new sources of capital, thus reducing public borrowing and improving the host government's credit rating;
- The ability to accelerate the development of projects that would otherwise have to wait for scarce sovereign resources;
- The use of private sector capital, initiative and know-how to reduce project construction costs and schedules and to improve operating efficiency;
- The allocation of project risk and burden to the private sector that would otherwise have to be undertaken by the public sector;
- The involvement of private sponsors and experienced commercial lenders, providing an in-depth review and additional assurance of project feasibility;
- Technology transfer, training of local personnel and development of national capital markets;
- In contrast to full privatisation, the government's retention of strategic control over the project, which is transferred back at the end of the contractual period; and
- The opportunity to establish a private benchmark to measure the efficiency of similar public sector projects and thereby offer opportunities for the enhancement of public management of infrastructure facilities.

Risk transfer is one of the main reasons for adopting the PPP approach. The private sector is in general more efficient in asset procurement and service delivery and as a result it is to the government's advantage to share the associated risks with the private sector. In line with widely accepted principles, the Hong Kong government's Efficiency Unit (2008) advocated that the most ideal situation is to allocate the risk to the party most able to manage and/or control that risk. For example, the contractor would take up the construction risk, the designer would take up the design risk, the government would take up environmental approval risks, land acquisition risks, and so on (Corbett and Smith 2006; Chan *et al.* 2006; Grimsey and Lewis 2004; Boussabaine 2007; Akintoye *et al.* 2003; Li *et al.* 2005a; Li 2003; Efficiency Unit 2008; Ingall 1997; New South Wales Government 2006; European Commission Directorate 2003; United Nations Economic Commission for Europe 2004; British Columbia 1999).

Cost certainty is more easily achieved in PPP projects as financial terms are identified and included within the contract. Since the private consortium will normally be responsible for financing, designing, constructing and operating the facility over an extended period, any cost saving can naturally result in a better chance of securing profit. Hence they are keen to control their spending tightly (Corbett and Smith 2006; Chan *et al.* 2006; Environment, Transport and Works Bureau 2004; Boussabaine 2007).

Innovation is another important advantage that the private sector can bring to public services. Generally speaking, the public sector may not be as innovative as the private sector. The private sector on the other hand is continuously searching for new products and services to increase their competitive edge and to save costs (Chan *et al.* 2006; Environment, Transport and Works Bureau 2004; Akintoye *et al.* 2003; Li *et al.* 2005b; Li 2003; Efficiency Unit 2008; New South Wales Government 2006; British Columbia 1999).

The private sector is made responsible for ensuring that the asset or service delivered meets pre-agreed quality benchmarks or standards throughout the life of the contract. Sometimes, the private consortium would only receive payment upon meeting certain requirements of the project; or it is motivated by the incentive payments to reward the high quality of service to be provided.

In a PPP project the consortium is also responsible for the long-term maintenance of the facility or service. The concession period may range from a few years to decades. Therefore, the consortium is keen to design and construct it to ensure better maintainability (Chan *et al.* 2006; Environment, Transport and Works Bureau 2004; Grimsey and Lewis 2004; Boussabaine 2007; Li 2003; Efficiency Unit 2008), at least within the concession period if not beyond.

Public sector projects delivered by the PPP model can often be completed on time and even with time savings because the consortium would start receiving revenue once the facilities/services are up and running. Therefore, the project team is keen to complete design and construction as quickly as possible. Once it starts to accrue revenue it can begin to pay off the initial costs and build up profits, whereas in a traditionally procured project there are no extra financial incentives for public servants to deliver projects faster. As a result, projects can at best proceed as scheduled (Environment, Transport and Works Bureau 2004; Grimsey and Lewis 2004; Akintoye *et al.* 2003; Li 2003; Efficiency Unit 2008).

Time certainty is found to be more easily achieved in PPP projects. The consortium is often paid according to milestones in the project schedule and any delay might be subject to liquidated damages. Therefore the consortium is often motivated to reach these milestones on time, if not earlier. This is a common behaviour observed in the private sector but it may not be the case in the public sector (Chan *et al.* 2006).

To the government, PPP frees up fiscal funds for other areas of public service, and improves cash flow management, as high upfront capital expenditure is replaced by periodic service payments and provides cost certainty in place of uncertain calls for asset maintenance and replacement. Public sector projects delivered via the private sector normally involve private sector funding. Consequently, the public funding required for public services can be reduced and redirected to support sectors of higher priority, such as education, healthcare, community services, and so on (Li *et al.* 2005b; Efficiency Unit 2012a).

To the private sector participants, PPP provides access to public sector markets. If priced accurately and if costs are managed effectively, the projects can provide reasonable profits and investment returns on a long-term basis. Also, these projects tend to be large and require expertise from many areas. Hence co-operation among different collaborating parties is encouraged (Environment, Transport and Works Bureau 2004; Grimsey and Lewis 2004; Boussabaine 2007; European Commission Directorate 2003; United Nations Economic Commission for Europe 2004).

Business opportunities are also created, due to the large scope of works that can benefit different sectors (Li 2003; Efficiency Unit 2008; United Nations Economic Commission for Europe 2004; British Columbia 1999).

Negative factors of PPP

Similarly the negative factors for PPP were also reviewed and a summary has been given in this section. Berg *et al.* (2002) also summarised some disadvantages of PPP projects:

- Lengthy bidding process: from initial phase of public sector assessment to signing of contract takes up to two years. The process of inviting, preparing, assessing and refining bids and negotiating contracts is complex and procedural.
- High bidding costs: the detailed and lengthy nature of the bidding process implies increased transaction costs.
- Small number of bidders
- Cost overruns: considerable scope for cost inflation through the bidding process.
- Excessive risks: not clear to what extent the government can shift risk.

The impact of risks to project objectives in completing a PPP project is usually significant, and these risks arise from multiple sources including the political, social, technical, economic and environmental factors, due mainly to the complexity and nature of the disciplines, public agencies and stakeholders involved. Both the private and public sectors need to have a better understanding of these risks in order to achieve an equitable risk allocation and enable the project to generate better outcomes (Chan *et al.* 2006; Environment, Transport and Works Bureau 2004; Gunnigan and Eaton 2006; Koppenjan 2005; Li 2003; Merna and Owen 1998; Mustafa 1999; Ng and Wong 2006; Satpathy and Das 2007; Xenidis and Angelides 2005; Zhang 2001; Zhang and AbouRisk 2006). In fact, a fair and reasonable allocation of various risks is vital to PPP success. If risks are inequitably or wrongly allocated beyond the capacity of the parties concerned, PPP projects would fail (for example, the demand risk resulting from town planning falling on a private consortium).

PPP projects may fall apart due to failure on the part of the private sector participants. In contracting out the PPP projects, the government should ensure that the parties in the private sector consortium are sufficiently competent and financially capable of taking up the projects. Due to a lack of relevant skills and experience of project partners, PPP projects are more complex to procure and implement (for example, the London Underground).

One common problem encountered in PPP projects is the high bidding costs, caused by increasing project complexity and protracted procurement process. The private sector incurs high bidding costs partly due to the consideration of the client's and their financiers' objectives. Lengthy negotiations and especially the cost of professional services may increase the bidding costs further (Chan *et al.* 2006; Corbett and Smith 2006; Environment, Transport and Works Bureau 2004; Li 2003; Li *et al.* 2005b; Mustafa 1999; Xenidis and Angelides 2005; Zhang 2001).

The PPP bidding process is also regarded as lengthy and complicated. For example, bidders are required to prepare tender proposals attached with a bundle of additional materials. Such a process may take three to four months. Besides, another several lengthy negotiations will be required for the formation of the contract. Clearly, setting up a complicated agreement framework for successful PPP implementation can slow down the bidding process (Chan *et al.* 2006; Environment, Transport and Works Bureau 2004; Grimsey and Lewis 2004; Li 2003; Li *et al.* 2005b; Merna and Owen 1998; Mustafa 1999; Zhang 2001).

One other reason for failure is opposition from the stakeholders and the general public. Whether the proposed project is consonant with the interest of the public is important, as public opposition can adversely affect the funding for the project from the public sector (El-Gohary *et al.* 2006; Grimsey and Lewis 2004; Zhang and AbouRisk 2006). PPP in public projects typically incurs political and social issues like land resumption, town planning, employment, heritage and environmental protection. These could result in public opposition, over-blown costs and delays to the projects.

Another common complaint by the public is the high tariff charged for the services provided. Often, the private sector faces a political uphill struggle in raising tariff to a level sufficient to cover its costs and earn reasonable profits and return on investment. The participation of the private sector in providing public service will undoubtedly bring innovations and efficiencies in the operation, but may produce a fear of downsizing in the public sector. To a certain extent, there would be fewer employment opportunities if no regulatory measures were implemented (Li 2003; Li *et al.* 2005b; Zhang and AbouRisk 2006).

The introduction of PPP exerts unprecedented pressure on the legal framework as it plays an important role in economic development, regeneration and mechanism for developing infrastructure. However, some countries do not have a well-established legal framework for PPP projects and the current legal framework is only supposed to deal with the traditional command and control model. Although PPP involves a great deal of legal structuring and documentation to deal with potential disputes amongst PPP parties, a 'watertight' legal framework is still lacking (for example, protection of public interests versus the legitimate rights of the private sector). Without a well-established legal framework, disputes are inevitable (Grimsey and Lewis 2004; Li *et al.* 2005b; Satpathy and Das 2007).

Private sector investors bear financial risks in funding the investment. Seeking financially strong partners in a PPP project is regarded as difficult. In most PPP arrangements, the debt is limited-recourse or non-recourse, where financiers need to bear risks. In fact, most stakeholders are not willing to accept excessive risks. The lack of mature financial engineering techniques on the part of the host countries can be another problem (Grimsey and Lewis 2004; Zhang 2001). An unattractive financial market (for example, with political instability or high interest rate) is often a negative factor to PPP success. Therefore, a conducive financial market is important for the private parties to drive PPP projects.

Value for money of PPP

One of the main reasons that projects are procured by PPP is to enhance value for money (VFM) by inviting the private sector to handle public works projects. As a result there has been much literature on how VFM in PPP projects can be achieved. This section reports only a few examples of how VFM can be achieved in PPP projects.

VFM, was defined by Grimsey and Lewis (2004) as the optimum combination of whole life cycle costs, risks, completion time and quality in order to meet public requirements, is another important consideration when deciding whether to proceed with the PPP option, especially for the public sector (Chan *et al.* 2006; Boussabaine 2007; Li *et al.* 2005b; Li 2003; Efficiency Unit 2008; Ingall 1997; New South Wales Government 2006; European Commission Directorate 2003). 'Public sector comparator' is the most common tool used by the public sector to show how much it would cost the government to build the asset through public funding, which is then used to compare with how much it would cost to build it as a PPP (Farrah 2007). In the case of the University College London Hospital Redevelopment in the UK, the PPP option cost 6.7 per cent less than the public sector comparator, while maintaining the same output and user requirements as demanded (Efficiency Unit 2012b).

Cost savings refer to the reduction in price as a result of delivering a project by PPP instead of traditional methods. The saving could be a result of the private sector's innovation and efficiency which the public sector may not be able to achieve (Corbett and Smith 2006; Environment, Transport and Works Bureau 2004; Grimsey and Lewis 2004; Akintoye *et al.* 2003; Li *et al.* 2005b; Li 2003; Efficiency Unit 2008; European Commission Directorate 2003; United Nations Economic Commission for Europe 2004; British Columbia 1999). The private sector generally achieves higher operational efficiency in asset procurement and service delivery by applying their expertise, experience, innovative ideas and/ or technology (such as using durable materials to reduce future maintenance cost) and continuous improvements. Overall cost savings to the project can be achieved by striving for the lowest possible total life cycle costs while maximising profits.

PPP project arrangements are complex and involve many parties with conflicting objectives and interests. Hence, PPP projects often require extensive expertise input and high costs, and deal negotiation can be lengthy. The high transaction costs and lengthy time may not represent good value to all parties and as a result the deal may not materialise in the beginning, or may falter in the end. PPP projects may incur higher transaction costs than those under the conventional public sector procurement. The legal and other advisory fees would be included as lawyers are involved in all stages of a PPP project, as well as the cost of private sector finance, and the price premium for single point responsibility arrangement. The potentially high transaction costs may have a negative impact on the objective of securing the best value (Corbett and Smith 2006; Environment, Transport and Works Bureau 2004; Grimsey and Lewis 2004; Li 2003; Li *et al.* 2005b; Merna and Owen 1998; Zhang 2001; Zhang and AbouRisk 2006). Complex PPP projects require inputs from many parties with differing expertise. Therefore, the projects should be economically viable to cover such costs.

Critical success factors of PPP

In order to achieve successful PPP projects, some suggestions have previously been reported in literature. This section reports only a few examples of how successful PPP projects can be achieved.

Under PPP contracts the government should be concerned that the assets are procured and services are delivered on time with good quality, and meet the pre-agreed service benchmarks or requirements throughout the life of the contract. However, the government should be less concerned with 'how' these are achieved and should not impose undue restrictions and constraints on the private sector participants. The government should be relegated to the primary role of industry and service regulation; it should be flexible in adopting innovations and new technology; it should provide strong support and make incentive payments to the private sector where appropriate. On the other hand, the government should retain control in case of default and be prepared to step in and re-provide the service if necessary (Abdul-Rashid *et al.* 2006; Corbett and Smith 2006; El-Gohary *et al.* 2006; Jamali 2004; Kanter 1999; Li *et al.* 2005c; Tam *et al.* 1994; Tiong 1996; Zhang 2005a).

A transparent and efficient procurement process is essential in lowering the transaction costs, shortening the time in negotiation, and completing the deal. Having a clear brief on the project and client requirements should help to achieve these in the bidding process. In most cases, competitive bidding solely on price may not help to secure a strong private consortium and obtain value for money for the public. The government should take a long-term view in seeking the right partner (Corbett and Smith 2006; Gentry and Fernandez 1997; Jefferies *et al.* 2002; Jefferies 2006; Li *et al.* 2005c; Qiao *et al.* 2001; Zhang 2005a).

Successful PPP implementation requires a stable political and social environment, which in turn relies on the stability and capability of the host government (Wong 2007). Political and social issues that go beyond the private sector's domain should be handled by the government. If unduly victimised, it is legitimate that the private sector participants should be adequately compensated. Unstable political and social environments have resulted in some failed rail projects (for example, frequent change in government premiers in Bangkok leading to the cancellation of many new public infrastructure projects originally procured under the PPP approach (Khang 1998; Cobb 2005)).

Many researchers (Akintoye *et al.* 2001; Corbett and Smith 2006; Jefferies *et al.* 2002; Li *et al.* 2005c, Zhang 2005a) have found that project financing is a key success factor for private sector investment in public infrastructure projects. The availability of an efficient and mature financial market with the benefits of low financing costs and a diversified range of financial products would be an incentive for private sector take-up of PPP projects.

Chapter summary

This chapter has provided some background on PPP projects in general. The traditional practice of procuring public works projects was reviewed to highlight the similarities and differences compared with the PPP method. The early developments of PPP have been briefly reviewed and the features of PPP presented. This chapter has formed an informative foundation for the following chapters in this book.

2 Financial models for public works projects

Introduction

There are many different types of financial models for delivering public works projects. This chapter defines these models broadly into six types according to the degree of involvement between the public and private sectors including: (1) public ownership and operation; (2) public funding with private operation; (3) joint ventures; (4) concession awards; (5) privatisation of state ownership; and (6) full private ownership and operation (Chan *et al.* 2007a). These models are explained according to the varying levels of involvement between the public and private sectors with the most public sector and least private sector involvement first, and the most private sector and least public sector involvement last. The selection of the most suited approach would very much depend on the project and also the priorities of the government involved.

Public ownership and operation

Public ownership and operation is the most common and traditional delivery mode for public projects. It is also commonly used for transportation type projects because of their high capital cost, high risk exposure and long-term operation. This mode is often adopted by the government for public projects which are in high demand by or of necessity to the general public. Under such conditions it is simpler for public ownership to deliver the projects speedily and avoid the lengthy procurement stage when private parties are involved. In public ownership and operation the government is not necessarily involved directly with the project, instead they may simply inject capital into corporatised concerns as the sole or major shareholder and exercise commercial prudence in its business activities. This form of public ownership and operation is particularly common for power and water projects in China where, traditionally, all public projects were delivered this way.

Nevertheless, public ownership and operation has often been linked with inefficiency and poor value for money. This is largely due to the fact that governments around the world all operate in a similar manner, which is according to a set of rules. Their priority is to do no wrong. As a result they tend to be criticised for lacking of drive and innovation. Controversially, the private sector often delivers better public services and facilities as they possess the most important ingredient: a business mindset. In order to profit they will provide expertise, innovation, speed, value for money, leadership, management, and so on. It may seem strange that given the numerous advantages which can be provided by the private sector, there are still many governments which would rather get the job done themselves. For example, in the United States, support for public ownership became a means of preventing corruption amongst government officials (Glaeser 2001). Experience in the US showed that when the private and public sectors are doing business, there is a large opportunity for corruption. Governments have often been found to overpay or undercharge the private sector. Consequently, exclusive government ownership and operation can eliminate these problems.

Public funding with private operation

This mode would include a service contract and leasing arrangement. It is regarded as the simplest form of partnership between the public and private sectors. Under this arrangement, the private service provider is responsible for the operating, repairing and maintenance costs. The capital costs are met by the public sector. As for operation and maintenance, the public sector either pays the private operator an annual fee, or more often expects the private operator to be self-financing (Chan *et al.* 2007a). Generally, the private operator is not responsible for any new capital investment or replacement of any asset (Li 2003). The purpose of this arrangement is to introduce the private sector's efficient management and innovations into the public sector, whilst maintaining to a large extent public sector control.

Joint ventures

Joint ventures are usually adopted as a way of realising the commercial potential of which the public project is capable (Pretorius *et al.* 2008). This method requires the skills of both the public and private sectors. Joint ventures usually delineate clearly the contributions of each party together with the risks and rewards involved. The private party is usually chosen via a competitive bidding process.

This mode provides an opportunity for direct collaboration between the public and private partners. The partners can either form a new joint venture or assume joint ownership of an existing company (for example, the public sector sells part of its shares of an existing company to the private sector, as in the case of the Mass Transit Railway Corporation in Hong Kong). Under joint ventures, the public and private sectors hold co-responsibility and co-ownership for the delivery of services (Li 2003).

Trafford and Proctor (2006) conducted a research study which looked at the key characteristics for successful implementation of joint ventures between the public and private sectors. The results indicated that these included good communication, openness, effective planning, ethos and direction. Their study found that there was a lack of communication between different levels of management where people often misunderstood motives or assumed motives; hence good communication was identified as one of the key characteristics for successful joint ventures.

Openness was another characteristic identified; it was found that often there was a lack of openness and trust between partners. Furthermore, in their study they realised that there was a lack of plan or method to identify whether objectives have been achieved successfully, hence planning was identified as a characteristic too. Another characteristic was ethos, which relates to the cultural and social differences between the two parties. The final characteristic identified was direction, which refers to the strong drive and ability of the joint venture to lead in a focused and positive direction in order to deliver a successful project.

Bult-Spiering and Dewulf (2006) also summarised some conditions for project success based on an extensive analysis of the results of case studies. These included: a joint venture mode gives a public–private partnership (PPP) arrangement a high chance of becoming successful; a larger scope in a development project tends to increase its performance; clear, timely and transparent mapping of all costs, revenues and profitability aspects of a PPP project is a necessary precondition; clear insight into the planning of project parts, the risk profiles involved, and the ways in which actors are involved in different project parts, is critical for the good performance of an urban development project.

On the contrary, some of the pitfalls of joint ventures may be avoided by having a clearer understanding of the stakeholders' interests; being aware of changes amongst different stakeholders; selecting the right organisational structure early; being prepared to be flexible; ensuring stakeholders' objectives are met; and using a multi-function approach when defining a PPP project (Bult-Spiering and Dewulf 2006).

Concession awards

Under a concession award, the private sector is responsible for financing, constructing and operating a new service facility or substantially renewing an existing one (Li 2003). The private partner retains the ownership of the facility and obtains financial return by directly charging the end-users or receiving annual payments from the government for a period. At the end of the concession period, the private sector would transfer the project back to the government, often free of charge. Build Operate Transfer (BOT) is a type of concession award.

Concession awards are typically defined by four features including: (1) the contract between the public and private sector; (2) a concession period; (3) the concessionaire is responsible for all investments and development during the concession period; and (4) the concessionaire is paid directly by the end-user or by a service fee from the government (Guasch 2012).

Guasch (2012) listed some of the advantages of concession awards which include: allowing private participation in projects that cannot be privately owned; enabling competition between private parties; and enabling cost efficiency. The disadvantages include: complex designs and monitoring systems as they tend to be large-scale projects; difficulty in enforcing contracts; the need for public accounting in the case of poor commitment from the private sector; lack of investment incentive towards the end of the concession period; and inability for price adjustments during the concession period.

Privatisation of state ownership

Privatisation of state ownership is a common practice nowadays. Privatisation would consist of the sale of a state-owned asset either by auction, public stock offering, private negotiation, or outright grant to a private organisation that assumes operating responsibilities (Li 2003). An example is the Japanese railway privatisation of 1987. Japan National Railways was split into six private regional passenger companies, a nationwide freight carrier and other related businesses such as telecommunications. This became a reform example for railways in other countries (Obermauer 2012).

Privatisation can be used as a means to increase competition and increase efficiency. There can be different stages to privatisation. In full privatisation the public enterprise will change its legal form to private but all shares are still held by the state. In actual privatisation, all shares will be held by the private sector. The level of privatisation adopted also represents the level of public sector involvement that still exists.

Full private ownership and operation

Full private ownership and operation makes use of the private sector's expertise, management and financial support to the maximum extent. The main differences compared to other procurement models are that privately owned assets can be sold or transferred at any time, they are not for a limited concession period, and also there is little or no involvement from the public sector.

Compared with the privatisation model, full private ownership and operation occurs when the private sector forms a private company on its own initiative and then provides public services with a licence granted by the government. Project economic viability may be one of the most critical factors for successful implementation, as it may require a huge capital investment and life cycle costs. If the project alone does not provide an attractive business model, supplementary or complementary businesses are usually carried out, such as property, shopping centres, hotels and so on.

Chapter summary

This chapter has presented briefly six procurement models for delivering projects. These models range from 100 per cent public participation to 100 per cent private participation. The six modes show varying levels of partnership between the public and private sectors. For each model there are strengths and weaknesses, as presented in Table 2.1. There is no perfect rule for which model to adopt. But some characteristics may be more suited for a specific type of model. Furthermore, each project is unique so that thorough consideration should be given before a decision is made.

Table 2.1 Strengths and weaknesses of different procurement models

Model	Strengths	Weaknesses
(1) Public ownership and operation	Traditional approach so more familiar Many lessons learnt Speedy procurement stage Prevents corruption	Inefficiency Lack of value for money Lack of innovation and drive Lack of business mindset
(2) Public funding with private operation	Simplest form of PPP Financial stability provided by public sector Private sector's efficiency, management, and innovation Public sector retains control	Limited room for private sector's innovation, creativity or expansion
(3) Joint ventures	Risks and rewards of each party listed and shared Good communication Openness Effective planning Ethos and direction Clear, timely, and transparent mapping of costs	Stakeholders may not be aware of each other's interests or changes
(4) Concession awards	Allowing private participation in projects that cannot be privately owned Enabling competition between private parties Enabling cost efficiency	Complex designs and monitoring systems as they tend to be large-scale projects Difficulty in enforcing contracts The need for public accounting in the case of poor commitment from the private sector Lack of investment incentive towards end of concession period Inability for price adjustments during concession period
(5) Privatisation of state ownership	Increases competition and efficiency of public projects	Requires continuous monitoring by public sector
(6) Full private ownership and operation	Makes use of private sector's expertise, management, and financial support to maximum Project can be sold or transferred at anytime Project not for a limited concession period Little or no involvement from the public sector.	Requires huge capital costs and operation costs from private party Project must be financially viable

Source: Chan *et al.* 2007a; Glaeser 2001; Li 2003; Pretorius *et al.* 2008; Trafford and Proctor 2006; Bult-Spiering and Dewulf 2006; Guasch 2012; Obermauer 2012

3 The development of public–private partnership internationally

Introduction

Public-private partnership (PPP) has been used internationally in more than eighty-five countries as a procurement method for delivering public infrastructure (Regan *et al.* 2009). Its main characteristics include a competitive bidding process, appropriate balance of project risks, private sector innovation and expertise, and improved public services and facilities (Chan *et al.* 2009a). Nevertheless, the adoption of PPP can vary drastically depending on its geographical location, as differences in experiences, history, culture, economy, and so on will all affect projects. This chapter takes a look at the experience of adopting PPP in six continents in order to realise how their experiences with PPP projects vary.

The development of PPP in North and South America

PPP in Canada

PPP was encouraged vigorously in 2007 to deal with ageing public facilities and services in Canada. Better and more facilities and services were urgently required at that time. In 2007, it was estimated that a further C\$123 billion (approximately US\$125 billion on 11 December 2012, Yahoo! Finance 2012) would be required to meet the demand (Canadian Council for Public Private Partnerships 2012). Additionally, local governments were looking at ways to deal with project delays and cost overruns frequently faced by traditional procurement. From the early 1990s to 2011, over 150 PPP projects were delivered across Canada.

The PPP market in Canada can be defined as advanced, emerging and undeveloped for different provinces within the country (PPP Canada 2012a). Advanced PPP markets tend to be the larger provinces with larger infrastructure budgets, and have institutionalised PPP procurement. These provinces include British Colombia, Alberta, Ontario and Quebec. Emerging PPP markets tend to have a PPP policy framework, a focal point for general PPP advice, and less experience with PPP projects. These provinces include Saskatchewan, Manitoba, New Brunswick, Nova Scotia, PEI and the NWT. Undeveloped PPP markets tend to have limited knowledge regarding PPP and also a weak institutional and financial ability to consider PPP projects.

The experience of Canada has demonstrated that PPP is best used for large and complex projects where innovation can reduce lifetime costs and deliver better infrastructure. Most provinces require that projects should be over C$40 million (approximately US$41 million on 11 December 2012, Yahoo! Finance 2012) before being considered, and in experience most range from C$100 million (approximately US$101 million on 11 December 2012, Yahoo! Finance 2012) to over C$1 billion (approximately US$1 billion on 11 December 2012, Yahoo! Finance 2012) (Canadian Council for Public Private Partnerships 2012). Due to the size limitation, it is believed that PPP is a better procurement option in only up to 20 per cent of all public works projects. For those cases where PPP has been the suitable option, large cost savings have been demonstrated. For example, for the Autoroute 30 project south of Montreal, more than C$750 million (approximately US$760 million on 11 December 2012, Yahoo! Finance 2012) was saved by using PPP (PPP Canada 2012a).

With the emergence of more PPP projects, PPP Canada was set up and became operational in 2009 to oversee and monitor PPP projects. The body has an independent Board of Directors which reports through the Minister of Finance to Parliament (PPP Canada 2012b). The body is similar to other PPP units across the world in that ensuring value for money is one of their high priorities.

PPP in Brazil

PPP was introduced in Brazil as a means to overcome deteriorating infrastructure and the lack of public resources for improvement. In 2004, the Brazilian Federal Law 11079 defined PPP as a supported concession (Diário Oficial da República Federativa do Brasil 2004). The first PPP project implemented in Brazil was the fourth line of the Metro of São Paulo. The contract was signed in 2006 with a consortium led by one of the major private toll road concession groups in Latin America. This project has a concession period of thirty years and will oversee the operation of a 12.8km stretch of subway in São Paulo. The consortium's investment amounted to US$340 million.

The development of PPP in Africa

PPP in South Africa

South Africa has adopted PPP since 1997 when an inter-departmental task team within the South African Cabinet was established to develop a package of policy, legislative and institutional reforms to create an enabling environment for PPP projects (National Treasury of Republic of South Africa 2012). From 1997 to 2000 there were six pilot PPP projects conducted including the N3 and N4 toll roads, two maximum security prisons, two water municipalities, and a tourism project. From these projects a strategic framework was endorsed by the South African Cabinet in 1999 and the Public Finance Management Act was passed in 2000. Also in 2000, the PPP Unit was established with the National Treasury consisting of five staff from both the public and private sectors. The main responsibilities of the PPP Unit are to provide technical assistance to government departments and provide Treasury approvals during pre-contract phases (Burger 2012).

Up to 2013 the PPP Unit comprises seventeen staff that oversee projects from a large range of industries including health, accommodation, energy, education, water, budget support, transport, contract management, ICT, project development facility, tourism, business development, waste and international relations. Some of the main priorities of the PPP Unit include affordability, value for money and transfer of risks to the private sector.

From 2000 to 2006 a further twelve PPP projects were signed (Burger 2012). The development of PPP projects has been quite slow due to the local government's lack of experience.

PPP in Nigeria

Nigeria has recently been reported to be the third-fastest growing economy in the world (Udemezue 2012). This rapid economic development forces local government to address the problems related to hunger and poverty, high unemployment rates, power instability, poor healthcare services, poor water quality and facilities, traffic jams and a high crime rate. Similar to those of other developing countries, Nigeria's government is unable to finance the developments required and hence private financing becomes essential to facilitate the country's much needed development.

The adoption of PPP in Nigeria is still relatively new; approximately a dozen large-scale PPP projects have been conducted (Detail Commercial Solicitors 2012). Ibrahim *et al.* (2006) concluded in their study that the risk factors in Nigeria include the unstable government, the inadequate experience in PPP, and the non-availability of finance. Furthermore, they advocated that the majority of risks should be allocated to the private sector, but that the public sector should retain political and site acquisition risks, and both parties should share relationship-based risks.

Awodele *et al.*'s (2012) study showed that the performance of PPP projects in Nigeria to date has not been ideal. They suggested that the local government should develop a holistic framework for attracting private investors, the public officers should be trained to have a good understanding of PPP concepts, and risk management concepts should be integrated into the model. These recommendations are thought to improve and encourage future PPP projects in Nigeria.

The development of PPP in Europe

Europe has a long history in using the PPP model. PPP projects can be dated as far back as the 1800s during the railway construction boom in the UK (Grimsey and Lewis 2004). Over the past two decades there has been a rapid boom in PPP projects across Europe. Some statistics are presented in Table 3.1, which shows the number and value of PPP projects in Europe from 1990 to 2009 (Kappeler and Nemoz 2012). The total number and value of projects conducted during this period was 1,340 and €253,745 million respectively (approximately US$336,295 million on 11 December 2012, Yahoo! Finance 2012). The statistics show a steep rise in projects both in terms of number and value from 1990 to 2009. In 1990 there were only 2 projects valuing €1,387 million (approximately US$1,838 million on 11 December 2012, Yahoo! Finance 2012). In 2009 the number of projects had risen to 118 which is fifty-nine times more than there were two decades ago. The total value of these projects was €15,740 million (approximately US$20,860 million on 11 December 2012, Yahoo! Finance 2012) which is eleven times more than two decades ago. From 1990 to 1994, the statistics show that there were only a few projects each year adopting the PPP model. From 1995 to 2003, the statistics show that the number and value of projects was rising steadily. The years 2004 to 2007 showed the highest number and value of PPP projects during this period. And finally a slight decrease in projects was observed in the last two years.

Furthermore, Table 3.2 shows the share of number and value of projects for each country in Europe from 1990 to 2009 (Kappeler and Nemoz 2012). The figures show that the UK was the leader in delivering PPP projects for both quantity and value at 67.1 per cent and 52.5 per cent respectively. Runner-up was Spain with 10.1 per cent and 11.4 per cent respectively.

Table 3.1 Evolution of PPP projects in Europe 1990–2009

Year	Number of projects	Value of projects (in € million)
1990	2	1,387
1991	1	73
1992	3	610
1993	1	454
1994	3	1,148
1995	12	3,265
1996	26	8,488
1997	33	5,278
1998	66	19,972
1999	77	9,603
2000	97	15,019
2001	79	13,315
2002	82	17,436
2003	90	17,357
2004	125	16,880
2005	130	26,794
2006	144	27,129
2007	136	29,598
2008	115	24,198
2009	118	15,740
Total	1,340	253,745

Source: Kappeler and Nemoz 2012.

Note: €1 = US$1.33 (Yahoo! Finance 2012)

Table 3.2 Share of number and value of projects for each country in Europe during 1990–2009

	% of no. of projects	% of value of projects
Austria	0.2	0.5
Belgium	0.9	1.3
Bulgaria	0.1	0.1
Cyprus	0.2	0.3
Czech Republic	0.2	0.3
Germany	4.9	4.1
Denmark	0.1	0.0
Greece	1.0	5.5
Spain	10.1	11.4
Finland	0.1	0.2
France	5.4	5.3
Hungary	0.7	2.3
Ireland	1.3	1.6
Italy	2.4	3.3
Latvia	0.1	0.0
Malta	0.0	0.0
Netherlands	1.2	1.8
Poland	0.4	1.7
Portugal	3.1	7.0
Romania	0.1	0.0
Sweden	0.1	0.2
Slovakia	0.1	0.5
Slovenia	0.1	0.0
United Kingdom	67.1	52.5
Total	100	100

Source: Kappeler and Nemoz 2012.

The development of PPP in Asia

PPP in China

In China, PPP projects have been introduced since the late 1970s as a means to encourage the country's reform (Adams *et al.* 2006). With the increasing demand for more and better infrastructure, the Chinese government started to apply PPP schemes at large scale from the 1990s by introducing more foreign investment especially for water, power and road projects (Sachs *et al.* 2007). Although the PPP model may appear attractive for overcoming the large amount of infrastructure development currently being conducted in China, there is a need to structure the existing practices of PPP adopted in other countries to suit the local economic, financial, legal and regulatory environment. In order to do so there are many challenges which are foreseeable (Chen and Doloi 2008).

China has already had some experience with PPP projects. Some of the more successful cases include Line 4 of Beijing Metro, the Beijing National Stadium (also referred to as the Bird's Nest), the Olympic Water Park project, the first sewage treatment plant of Shanghai Zhuyuan, the Hangzhou Bay Bridge, Line 4 of Shenzhen Metro, the sewage treatment projects in Canton Xilang, and the ten water plants in Beijing. These cases have demonstrated that the PPP model is easier for financing in a shorter amount of time, reducing the financial burden on the local government, investment diversification, and providing a reasonable amount of risk sharing (Qu and Li 2009). Consequently, PPP can be seen as beneficial to ease the financial pressure on the Chinese government. In addition, as these projects are normally large scale, the profits are particularly attractive to the private sector. The win–win idea means that both the public and private parties are supportive of adopting the PPP arrangement for projects in China.

The Chinese government believes that PPP is an effective way to ease their financial burden (Liu and Yamamoto 2009). Furthermore, they also believe that it is more efficient than the traditional model of financing. Other achievable benefits include flexible management mechanisms, expertise and cost-awareness. However, the implementation of PPP in China requires certain conditions. For example, the investment system should be improved to facilitate further partnerships, and the policy and legal environment should be more mature.

PPP in Hong Kong

Hong Kong is not completely new to the idea of PPP. In fact the city was probably one of the first to utilise resources from the private sector. The term PPP may sound revolutionary to Hong Kong, where a more familiar term is Build Operate Transfer (BOT). The concept of BOT has been used since the late 1960s. Although Hong Kong has had experience in adopting quite a number of BOT projects, the approach of PPP has not been studied extensively in the local context. Overseas experience demonstrating the benefits of PPP has re-initiated the interest of some local governmental departments to develop the traditional BOT arrangement into a more appropriate, refined and internationally recognised successful approach. The traditional practice of these projects was for the government to directly award a concession to the potential bidder. This practice of awarding concessions is common in Hong Kong, but the gestation period spent in formulating the enabling legislation is lengthy (Zhang 2001). Hong Kong being the international gateway to China, and possibly even to Asia, represents a huge business market filled with opportunities and attractions. Because of the foreseeable profit, Hong Kong has the potential to draw companies from across the world. Money coming in from outside is beneficial to the local government. The local government having seen the success stories experienced by others is keen to bring innovation and efficiency into their public works projects (Smith 2012).

In recent years the Efficiency Unit of the Hong Kong government has been heavily involved in PPP research. The local governments' interest in utilising PPP is obvious. The approaches that they have taken mainly involve gaining international experience, from Europe and Australia particularly. One of the early documents produced by the Efficiency Unit on private sector involvement was a guideline to help governmental bureaus and departments to familiarise them with private sector engagement (Efficiency Unit 2001). These guidelines were published in 2001 and showed the government's interest in adopting the idea of PPP. Only two years later they also produced a comprehensive introductory guide to PPP (Efficiency Unit 2003). This guide was designed for the use of the civil servants but is also made available to the public to help understanding of the government's approach. After the publication of this report much interest was drawn from the construction industry because of the possible increase in business opportunities.

More recently, the Efficiency Unit published two more guidelines on PPP (Efficiency Unit 2007; 2008). The first of these publications shows how more knowledge on the issues of PPP has been gained; it also identifies areas of concern to local practitioners as well as to civil servants, and it tries to provide some insights into these areas. The second publication is much more specific on how to establish a PPP project. The guideline is aimed at coaching civil servants on how to conduct a PPP project by looking at the business case, dealing with the private sector, managing the risks, funding and payment issues, managing performance, and so on.

PPP in India

India experienced a dramatic change after gaining independence from British rule in 1947. Political influence restricted foreign investments and imports as a means of demonstrating their independence. But in the 1990s, India's debt situation and budget shortages finally led to a situation where it was proving too difficult to cut off external links completely (Pretorius *et al.* 2008). Consequently, two major changes, including allowing industrial licensing and foreign investment, were introduced in 1991. The power sector was one of the first opened up for private investors as it was critical to India's economic development. The country's power demand was increasing annually at 8 per cent; Hill (2005) estimated that the power demand in 2005 would almost double that in 1990. This urgent demand for power supply opened up new investment opportunities for the private sector, where they were invited to build and operate power plants with no restrictions on foreign ownership.

The development of PPP in Australia

PPP has been an increasingly popular choice for delivering public works projects in Australia. Although for decades there have been public works projects delivered in Australia by similar partnership arrangements, it has only been since the early 1990s that PPP was first properly introduced. PPP has been a growing alternative to procuring public projects across the world. Especially with the success seen from the state of Victoria, the other Australian states are eager to get a taste.

The practice for delivering public works projects across Australia is quite different depending on the state. Each state government will have its own set of guidelines, rules, preferences and practice to go by. Political decisions are crucial in deciding procurement processes.

PPP in Victoria

The Victoria government released the Partnerships Victoria policy in June 2000 providing a framework for developing contractual partnerships between the public and the private sectors for public infrastructure and services (Partnerships Victoria 2000). This brought about the change to the traditional practice of using Build Own Operate (BOO) and Build Own Operate Transfer (BOOT). The traditional practice focused more on the private sector's financial input and also having the risk transferred from the public sector to the private sector. But since the Partnerships Victoria policy the focus moved more towards delivering better projects as a result of bringing in the private sector expertise, and also the government would regain direct control over the service or facility after the concession period.

The Partnerships Victoria team is part of the Commercial Division in the Department of Treasury and Finance of the Victoria state. The team is mainly responsible for overseeing projects implemented via the PPP practice and also developing guidelines and policies for PPP projects. Up to 2008, seventeen projects have already been implemented under Partnerships Victoria totalling AU$5.5 billion (approximately US$5.7 billion on 11 December 2012, Yahoo! Finance 2012) (Partnerships Victoria 2008a). The team has also produced four policies, four guidelines, three technical notes, and four advisory notes for the implementation of PPP projects in Victoria. These publications are targeted for the use of both the public and private sectors, and cover areas including the public sector comparator, risk allocation, standard commercial principles, tender process, interest rates, and so on (Partnerships Victoria 2008b).

PPP in Queensland

In the past, Queensland has always been a conservative state (Townsend 2004). The conservativeness of the Queensland government has meant that alternative procurement methods have not been considered until recently. Queensland is a unique state; for example it is the only state that has elected a communist member into its parliament. Another example more relevant to project procurement is that unlike other states such as New South Wales and Victoria, Queensland has its own public works department to deal with its projects. In other states those services tend to be outsourced to the private sector. The Queensland government still runs their own toll roads and some of the electricity. This may also be one of the reasons that other states have been more comfortable with PPP and quicker in its adoption.

Another major factor is that Queensland has never had a huge budgetary crisis. The traditional practice of the state operation is to retain the ability to control almost every function within itself, such as marketing functions, agricultural functions, and so on. The approach in general is therefore a very socialist one. Queensland has therefore adopted PPP much more slowly than other states in Australia. Victoria, for example, introduced PPP projects over a decade earlier.

PPP (although under a different name) was first utilised twenty years ago in Queensland, even before other states. The project was a toll way built along the Sunshine Coast. The short-lived procurement alternative came to a halt after the Queensland government changed political leadership from Conservative to Labour. As a result, public infrastructure was in essence nationalised. This was a typical intervention. The new Queensland government was not keen on adopting PPP. In other jurisdictions PPP projects have been driven by either the Premier or the Treasurer. And the Queensland Treasurer did not take any action to encourage PPP projects (Chan *et al.* 2008a).

Another major reason for which the Queensland government was not keen to jump into using PPP at first, is because it does not see the delivery method as value for money when compared to conventional options. The Queensland government's state budget is in a much more robust situation compared to that of Victoria when it first adopted PPP. Therefore there has never been any economic pressure for it to try out alternatives. Similar to other governments around the world, PPP was often first adopted due to budgetary crisis for delivering public infrastructure and services.

Queensland government's recent interest in PPP has also been due to changes in their economic situation. The state is required to pay approximately AU\$50 billion (approximately US\$52 billion on 11 December 2012, Yahoo! Finance 2012) over the next four years to uplift their infrastructure (Chan *et al.* 2008a). Therefore, there has been a lot more pressure in the state to find the money easily. PPP has therefore become an attractive option for drawing in cash from the private sector. Obviously, with the amount of experience and research done on PPP in other parts of Australia, the Queensland government is aware that PPP should be driven by value for money. The reason there have not been many PPP projects in Queensland is because some people in the state government believe that they may not represent value for money (Chan *et al.* 2008a). They see no reason for the private sector to finance and build public projects when they can borrow the money themselves at a lower financial cost.

The Queensland government has also been careful to ensure that it does not follow in the footsteps of the Sydney government. Although the Sydney government has delivered many successful PPP projects in the past, a very high-profile project has caused negative views on PPP. The Cross City Tunnel faced problems associated with traffic forecasts and toll fares (Chan *et al.* 2008b). The result was that the private consortium made a total loss. The risks associated with PPP projects were therefore highly profiled at the time causing the government unwanted media attention. Governments in general are concerned about similar criticism. Therefore the Queensland government has been careful with every step they take on the PPP path in order to avoid such traps. (The Cross City Tunnel case is discussed in Chapter 10.)

Many of the Queensland government's PPP guidelines have been based on those of neighbouring states, especially Victoria. Victoria's PPP process published by Partnerships Victoria (2000) has been undoubtedly the most advanced in Australia. The Victoria government has handled the most PPP projects in Australia in terms of number and variation. The publications produced are renowned worldwide and compared to those of Partnerships UK. Many players in the private sector have also tended to follow the rules in these publications, as often what is written by the Victorian government on PPP is believed to be the 'Bible of PPP' across Australia. Undoubtedly, the private sector may not even challenge the appropriateness of the guidelines, but often governments will expect the private sector to go by these 'rules', as it has become the expected norm. The Queensland government has moved one step forward by producing its own guidelines for PPP projects. Although based on the Victoria model, it has added some further steps in the process to suit Queensland's situation and needs.

Chapter summary

PPP has become a highly popular approach for procuring public works projects. This chapter has highlighted some of these developments in countries which have adopted this approach successfully and innovatively. It is interesting to see how these projects can vary tremendously between different places around the world. This has shown that there is no unique, proper, or correct way for adopting the PPP method: instead it should be shaped to fit the needs, purpose and practice of the country or even jurisdiction for which it is used.

Part II

Practitioners' perspective on public–private partnership

4 Different perspectives on procuring public works projects

Introduction

Interviews were conducted with thirty-five experts from the public sector, private sector and researchers in order to extract expert knowledge regarding the adoption of public–private partnership (PPP) from different perspectives in Hong Kong and Australia (Chan *et al.* 2008c; 2008d; 2008e; Cheung *et al.* 2010a; 2010b). The findings from these interviews were analysed according to the perspectives of the public sector, private sector and researchers. Within each of these categories, the findings from Hong Kong and Australian interviewees were further compared.

The public sector's perspective

Selecting respondents

The target public sector respondents of the interviews were practitioners of senior level and authority who have had experience in PPP. A total of fourteen interviews were conducted with experts from the public sector. Seven interviews were conducted in each jurisdiction. Amongst the seven interviews conducted in Hong Kong, two were from Administration Departments (one of the interviewees previously represented a Works Department), three were from Works Departments (one of whom previously represented an Administration Department while the other also holds a position at a local institute), two of the interviewees were from non-governmental organisations (NGO) (both had previously acted for different Works Departments).

The Australian interviewees comprised three government officials and four specialist advisers from the private sector. The government officials interviewed were from local state education and treasury departments. When arranging the interviews in Australia, it was found that the state governments tended to employ advisers from the private sector to act on their behalf in providing advice and expertise for selecting and monitoring the PPP project consortia. Therefore four advisers from the private sector were also selected for interview. Their roles were solely on behalf of the public sector, and so their responses can also be regarded as the public sector's view. Background details of these experts are shown in Tables 4.1 and 4.2 for Hong Kong and Australian interviewees respectively.

Table 4.1 List of interviewees from the public sector in Hong Kong

No.	Position of interviewee	Organisation of interviewee	Experience of interviewee
PU1	Assistant Director	Administration Department	Produced many PPP guidelines and conducted research
PU2	Permanent Secretary	Administration Bureau (previously Works Department)	Involved with initiating PPP projects in a works department he previously worked for
PU3	Director	Works Department (previously Administration Bureau)	Involved with ongoing PPP projects
PU4	Assistant Director	Works Department	Involved with ongoing PPP projects
PU5	Senior Quantity Surveyor	Works Department/ Local Professional Institute	Founder of a PPP research working group for his institute
PU6	Executive Board Member	NGO (previously Works Department)	Involved with initiating PPP projects in a works department he previously worked for
PU7	Executive Director	NGO (previously Works Department)	Involved with initiating PPP projects in a works department he previously worked for

Source: Cheung *et al.* 2010a; Chan *et al.* 2008c; 2008d (with permission from Taylor & Francis Group, Professor Sarosh Hashmat Lodi, Editor of *Proceedings of the First International Conference on Construction in Developing Countries*, Pakistan, and International Council for Research and Innovation in Building and Construction).

Interview findings

Table 4.3 shows a summary of the responses to each question given by the fourteen interviewees. The number of times that each response was given was tallied. Where the response was only given once it was believed to be insignificant for further analysis. For the responses given more than once, these were tabulated and further analysed as shown in Tables 4.4 to 4.9. The numbers in brackets represent the number of times the response was mentioned by interviewees.

Research on local case studies

Table 4.4 shows the responses to Question 1: 'Have you conducted any research looking at local case studies?' that were given more than once. The findings show that three different responses were given by Hong Kong interviewees and four were given by the Australians. Amongst the four responses given by the Australian interviewees, three were the same as those given by the Hong Kong interviewees. The response which was given most by both groups of interviewees was 'Other research conducted', mentioned five times for each. This finding showed that irrespective of geographical locations the interviewees tended to conduct other research besides case studies on PPP.

Table 4.2 List of interviewees from the public sector in Australia

No.	Position of interviewee	Organisation of interviewee	Experience of interviewee
PU8	Executive Director	Education Department	His department initiated a social PPP project of which he was responsible for the overall delivery
PU9	Director	Treasury Department	Involved with delivering many PPP projects, producing guidelines, training, courses and research
PU10	Executive Manager	Treasury Department	Involved with delivering many PPP projects, producing guidelines, training, courses and research
PU11	Executive Director	Transaction Adviser	Acted as government's adviser for many PPP projects, managing the tender, evaluation, negotiation and award
PU12	Partner	Legal Adviser	Acted as the government's legal adviser for many PPP projects
PU13	Head	Finance Adviser	Acted as the government's financial adviser for many PPP projects
PU14	Director	Finance Adviser	Acted as the government's financial adviser for many PPP projects

Source: Cheung *et al.* 2010a (with permission from Taylor & Francis Group).

The response 'Local case studies' was mentioned four times by the Australians. It is possible that because Australia has had much experience in conducting PPP projects, they do not need to look elsewhere to learn from the experience of others, instead they can refer to their own projects as reference material. As mentioned previously, the Australian state of Victoria for example has a large range of guidance materials on the public domain which other states can refer to when conducting PPP projects (Partnerships Victoria 2008b).

On the other hand, the Hong Kong interviewees mentioned 'International case studies' three times showing their need to learn from the experience of others. The Efficiency Unit of the Hong Kong government has also been known to be interested in international case studies. They have published a number of case study reports for PPP projects in the UK and Australia (Efficiency Unit 2012c). The Australians also mentioned this response two times. From the interviews it was found that the involvement in research was 'Not mentioned' twice by each group of interviewees.

Comparing PPP with traditional procurement methods

Table 4.5 shows the responses mentioned more than once by both groups of interviewees for Question 2: 'How would you compare PPP with traditional procurement methods?' Three and two different responses were mentioned more than

Table 4.3 Summary of responses from public sector interviewees

	Hong Kong interviewees								Australian interviewees							
	PU1	PU2	PU3	PU4	PU5	PU6	PU7	Total	PU8	PU9	PU10	PU11	PU12	PU13	PU14	Total
1. Have you conducted any research looking at local case studies? And if so, could you share your insights?																
Local case studies	✓							1		✓	✓			✓	✓	4
International case studies		✓	✓					3						✓	✓	2
Other research conducted		✓	✓	✓	✓			5	✓	✓	✓			✓	✓	5
Not mentioned						✓	✓	2				✓	✓			2
2. How would you compare PPP with traditional procurement methods?																
Using a public sector comparator	✓	✓						2								0
Longer tendering/negotiation for PPP	✓							1	✓							1
Government act as supervisor in PPP		✓						1								0
Traditional method accepted as norm			✓					1								0
Each project is unique		✓	✓					2								0
Difference in payment mechanism				✓				1								0
PPP projects gain private sector's added efficiency/expertise/management skills				✓		✓		2				✓			✓	2
PPP projects delivered faster					✓			1								0
PPP utilises private sector finance/difference in finance structure						✓		1			✓	✓		✓		4
PPP tend to be large project sums						✓		1								0
Difference in risk profile								0	✓							1

| | Hong Kong interviewees | | | | | | | | Australian interviewees | | | | | | | |
	PU1	PU2	PU3	PU4	PU5	PU6	PU7	Total	PU8	PU9	PU10	PU11	PU12	PU13	PU14	Total
Operational differences								0	✓							1
Management differences								0	✓							1
PPP projects have a more transparent process								0		✓						1
PPP considers whole life cycle cost								0			✓					1
More parties involved in PPP								0					✓			1
3. Which type of project do you feel is best suited to use PPP?																
Link between performance and payment	✓							1								0
Each project unique		✓						1								0
Economically viable			✓		✓	✓		3		✓				✓	✓	3
Value for money			✓					1								0
Large operating element/cost				✓		✓		2								0
Performance easily measured				✓				1		✓					✓	2
Mutual benefits for all parties							✓	1								0
Economic infrastructure								0	✓							1
Scope for innovation								0		✓				✓	✓	3
High project costs								0			✓	✓				2
Any nature								0				✓	✓			2
Sufficient risk transfer								0						✓	✓	2

Continued overleaf

	Hong Kong interviewees								Australian interviewees							
	PU1	PU2	PU3	PU4	PU5	PU6	PU7	Total	PU8	PU9	PU10	PU11	PU12	PU13	PU14	Total
4. What do you feel are the key performance indicators in a PPP project?																
Project performance	✓							1	✓	✓						2
Resources saved		✓	✓					2								0
Contractor's performance			✓	✓				2	✓				✓			2
Traditional KPIs: cost, time, quality					✓	✓	✓	3	✓							1
Risk management						✓		1	✓	✓						2
Public acceptance							✓	1	✓							0
Value for money achieved								0	✓							1
Service outcomes								0			✓					1
Contract terms								0				✓	✓	✓	✓	4
Client satisfaction								0					✓			1
Payment mechanism performed								0						✓		1
5. In general, what do you think are the critical success factors leading to successful PPP projects?																
Champion	✓							1		✓	✓				✓	3
Large project capital value	✓							1								0
Well prepared contract/document		✓		✓				2						✓		1
Partnership spirit/commitment/trust		✓				✓		2	✓						✓	2
Transparent process		✓				✓		2	✓							1
Project objectives well defined		✓		✓			✓	3	✓	✓	✓					3

	Hong Kong interviewees								Australian interviewees							
	PU1	PU2	PU3	PU4	PU5	PU6	PU7	Total	PU8	PU9	PU10	PU11	PU12	PU13	PU14	Total
Public consultation		✓			✓	✓		3				✓				1
Appropriate risk allocation			✓		✓	✓	✓	4				✓				1
Large operating element				✓				1								0
Development potential					✓			1								0
Economically viable						✓		1				✓		✓		2
Effective negotiations between parties								0	✓							1
Competitive procurement process								0	✓	✓	✓	✓	✓			5
Government support								0	✓					✓		2
Skilled and experienced parties								0		✓	✓	✓		✓		4
Clear milestones								0		✓	✓	✓				3
Initiate project								0				✓				1
Value for money								0				✓			✓	2
6. Does your organisation have any in-house guidance/practice notes?																
Yes	✓				✓	✓		3		✓	✓	✓	✓	✓	✓	6
No		✓	✓	✓			✓	4								0
Refer to others			✓	✓				2	✓							1

Source: Cheung *et al.* 2010a (with permission from Taylor & Francis Group).

Table 4.4 Question 1: 'Have you conducted any research looking at local case studies?'

Hong Kong interviewees	Australian interviewees
Other research conducted (5)	Other research conducted (5)
International case studies (3)	Local case studies (4)
Not mentioned (2)	Not mentioned (2)
	International case studies (2)

Source: Cheung *et al.* 2010a (with permission from Taylor & Francis Group).

Table 4.5 Question 2: 'How would you compare PPP with traditional procurement methods?'

Hong Kong interviewees	Australian interviewees
Using a public sector comparator (2)	PPP utilises private sector finance/ difference in finance structure (4)
PPP projects gain private sector's added efficiency/expertise/management skills (2)	PPP projects gain private sector's added efficiency/expertise/management skills (2)
Each project is unique (2)	

Source: Cheung *et al.* 2010a (with permission from Taylor & Francis Group).

once by the Hong Kong and Australian interviewees respectively. For all three responses mentioned by the Hong Kong interviewees each was mentioned twice.

Mentioned the most by Australian interviewees was 'PPP utilises private sector finance/difference in finance structure' which was mentioned four times. This finding shows the importance of the different financing structure provided by PPP projects. Although finance should not be the main reason for adopting PPP projects, undoubtedly, financial drive is still an attractive factor to governments, hence this response was unsurprising.

Mentioned by both groups of interviewees was the response 'PPP projects gain private sector's added efficiency/expertise/management skills'. This response was also mentioned twice by the Australian interviewees. From previous literature it has also been recorded that one of the main advantages of involving the private sector is to add value to public projects in terms of their efficiency, expertise and management skills when compared to those of the public sector (Yescombe 2008; Carrillo *et al.* 2007; Leiringer 2006).

Other responses mentioned by the Hong Kong interviewees included 'Using a public sector comparator', which was also mentioned by the Efficiency Unit (2003) of the Hong Kong government as necessary whenever public money is involved. Also 'Each project is unique' was mentioned by the Hong Kong interviewees too.

Projects best suited to use PPP

The interviewees were asked to answer 'Which type of project do you feel is best suited to use PPP?' in Question 3. Table 4.6 shows their responses that were mentioned more than once. The results showed that only one similar response was mentioned by both groups of interviewees. This was 'Economically viable' which was mentioned three times by both groups of interviewees and also mentioned the most. The private sector parties are business people, so for them to participate in PPP projects they must foresee reasonable financial benefits. Partnerships Victoria (2000) explains how developing a business case is a key step in the decision-making process. This is where the project is fully scoped and the risks and costs are identified to develop a cost–benefit analysis, as well as to test the net benefit of the proposal.

The Hong Kong interviewees suggested only one more criterion for PPP projects, which was 'Large operating element/cost' mentioned twice. One typical feature of PPP projects is that the consortium is normally responsible for the operation and maintenance of the project. Without this element PPP projects would be similar to projects procured traditionally. Therefore the operation part must constitute a reasonable proportion of the project. Grimsey and Lewis (2004) listed a number of public–private business models prior to the more general term PPP, many of which emphasised the operation element of the structure within their name, showing its highly important role in these arrangements. They include: Operate and Maintain (O&M); Operate Maintain and Manage (OM&M); Build Transfer Operate (BTO); Build Operate Transfer (BOT); Build Own Operate Remove (BOOR); Build Own Operate Transfer (BOOT); Lease Renovate Operate Transfer (LROT); Design Build Finance Operate (DBFO); Design Build Finance Operate Manage (DBFOM); Build Own Operate (BOO), and so on.

Other responses given by the Australian interviewees included 'Scope for innovation' (Eaton *et al.* 2006) which was mentioned three times. Also mentioned twice each by the Australians were 'Performance easily measured' (Partnerships Victoria 2000), 'High project value' (HM Treasury 2003), 'Any nature' and 'Sufficient risk transfer' (Jin and Doloi 2008). These features of suitable PPP projects have been previously recorded by other researchers as well.

Table 4.6 Question 3: 'Which type of project do you feel is best to use PPP?'

Hong Kong interviewees	Australian interviewees
Economically viable (3)	Economically viable (3)
Large operating element/cost (2)	Scope for innovation (3)
	Performance easily measured (2)
	High project value (2)
	Any nature (2)
	Sufficient risk transfer (2)

Source: Cheung *et al.* 2010a (with permission from Taylor & Francis Group).

Key performance indicators in PPP projects

The interviewees were also asked to answer Question 4: 'What do you feel are the key performance indicators in a PPP project?' (Table 4.7). Amongst the responses received, three were mentioned more than once by the Hong Kong interviewees and four by the Australian interviewees.

The response 'Contract terms' was mentioned the most at four times by the Australian interviewees. In Australia high priority is given to the contract component of projects procured by PPP. Guidelines have also been published on this aspect (Partnerships Victoria 2008b). The response mentioned the most by Hong Kong interviewees was 'Traditional KPIs [key performance indicators]: cost, time, quality'. Probably due to the lack of experience in PPP projects (not including BOT-type projects), the Hong Kong interviewees did not commonly come up with any responses that were specifically related to PPP projects. Only one response was raised by both groups of interviewees, which was 'Contractor's performance' which was mentioned twice by each group of interviewees.

Also mentioned twice by the Australian interviewees were the responses 'Project performance' and 'Risk management'. The performance of the contractor and project are items which would definitely be mentioned in the contract documents; these again confirm the importance of the contract to the Australian interviewees. Many studies have been conducted on the importance of risks in PPP projects (Akbiyikli and Eaton 2004; Li *et al.* 2004; Li *et al.* 2005a; Shen and Wu 2005). One of the main reasons for implementing public projects by PPP is also for risk transfer, therefore to classify the risk management as a performance indicator is also reasonable.

Another response mentioned by Hong Kong interviewees was 'Resources saved'. PPP projects are normally only conducted after they have been proved to be a cheaper alternative to traditionally procured projects. This is normally conducted via the public sector comparator (Efficiency Unit 2003; Partnerships Victoria 2008b).

Table 4.7 Question 4: 'What do you feel are the key performance indicators in a PPP project?'

Hong Kong interviewees	Australian interviewees
Traditional KPIs: cost, time, quality (3)	Contract terms (4)
Contractor's performance (2)	Contractor's performance (2)
Resources saved (2)	Project performance (2)
	Risk management (2)

Source: Cheung *et al.* 2010a (with permission from Taylor & Francis Group).

Critical success factors leading to successful PPP projects

Question 5: 'In general, what do you think are the critical success factors leading to successful PPP projects?' received the most variation of responses from the interviewees (Table 4.8). This probably indicated that there are many ways for PPP projects to achieve success.

For responses that were mentioned more than once, there were six from the Hong Kong interviewees and nine for the Australian interviewees. Amongst these, only two were similar for both groups of interviewees; these included 'Project objectives well defined' which was mentioned three times by each group of respondents, and 'Partnership spirit/commitment/trust' mentioned twice by each group of interviewees. As mentioned by the Efficiency Unit (2008) and the Queensland Government (2008a) the objectives/output specification of a PPP project must be well defined. The importance of partnership spirit was also identified by Gunnigan and Eaton (2006).

Mentioned the most frequently by Australian interviewees was 'Competitive procurement process' (Jefferies *et al.* 2002) at five times, followed by 'Skilled and experienced parties' (Drew 2005) at four times, 'Champion' (Efficiency Unit 2008) and 'Clear milestones' (Civic Exchange *et al.* 2005) both three times and 'Economically viable' (Chege 2001), 'Government support' (Qiao *et al.* 2001) and 'Value for money' (Heald 2003) all twice.

Mentioned the most by Hong Kong interviewees was 'Appropriate risk allocation' (Li *et al.* 2005a) at four times, 'Public consultation' (Kanakoudis *et al.* 2007) at three times and 'Well prepared contract/document' (Partnerships Victoria 2008c) and 'Transparent process' (United Nations Economic Commission for Europe 2004) both at two times.

Table 4.8 Question 5: 'In general, what do you think are the critical success factors leading to successful PPP projects?'

Hong Kong interviewees	Australian interviewees
Appropriate risk allocation (4)	Competitive procurement process (5)
Public consultation (3)	Skilled and experienced parties (4)
Project objectives well defined (3)	Project objectives well defined (3)
Well prepared contract/document (2)	Champion (3)
Transparent process (2)	Clear milestones (3)
Partnership spirit/commitment/trust (2)	Partnership spirit/commitment/trust (2)
	Economically viable (2)
	Government support (2)
	Value for money (2)

Source: Cheung *et al.* 2010a (with permission from Taylor & Francis Group).

In-house guidance/practice notes

For Question 6: 'Does your organisation have any in-house guidance/practice notes?' it was found that the majority of the interviewees (six out of seven) in Australia responded 'Yes', whereas only three interviewees in Hong Kong agreed (Table 4.9).

Four Hong Kong interviewees responded 'No' and two responded 'Refer to others'. This finding has shown that the Australians were much more likely to have their own guidance materials, whereas for the Hong Kong interviewees the responses varied. Australia has implemented many more PPP projects compared to Hong Kong; hence they can also be regarded as much more experienced. The Australian state of Victoria alone has implemented seventeen projects under the Partnerships Victoria arrangement (Partnerships Victoria 2008a) as mentioned previously. On the other hand, not considering the previous projects conducted by BOT, Hong Kong has only completed a couple of PPP projects.

Summary of findings from the public sector

This section has studied the public sector's perspective on procuring public works projects via findings from fourteen interviews conducted in Hong Kong and Australia. Government officials and advisers with experience in PPP projects and research were invited to answer six questions related to the implementation of PPP.

The results found that interviewees from both jurisdictions had conducted some kind of research in the area and had looked at international cases. This finding has shown that governments in both jurisdictions have shown an interest in other sources of information besides real cases and also both are keen to learn from international experiences. Therefore other governments can also consider using a similar approach if they have not already done so.

Both groups of interviewees also found that the main difference between PPP and traditional projects is that in a PPP project there is the added advantage of the private sector's efficiency, expertise and management skills involved. Therefore other governments could consider whether this added advantage is required from the private sector when they consider whether or not to opt for the PPP model in their public work projects. The interviewees from Hong Kong also suggested using the public sector comparator as an indicator to determine the preference between the methods. Other criteria recommended by the Australian interviewees were the private sector financing and finance structure of the project. Again these could be used as indications of which method to opt for.

Table 4.9 Question 6: 'Does your organisation have any in-house guidance/practice notes?'

Hong Kong interviewees	Australian interviewees
No (4)	Yes (6)
Yes (3)	
Refer to others (2)	

The interviewees were asked which projects would be suitable to use PPP; both groups suggested that an economically viable project would be crucial. Another important feature according to the Australian interviewees is scope for innovation.

It was suggested by both groups of interviewees that the contractor's performance would be the key performance indicator in a PPP project. The Hong Kong interviewees also suggested that the traditional key performance indicators such as cost, time and quality are also important. The Australian interviewees suggested that the contract terms should be considered. These findings are valuable for measuring the performance of a PPP project for both the public and private sectors.

Common critical success factors mentioned by both groups of interviewees included the project objectives being well defined and a partnering spirit, commitment and trust. These factors should be considered by all parties before the project begins to ensure that they are achieved. The Hong Kong interviewees also felt strongly that an appropriate risk allocation would achieve success in the project. For the Australian interviewees a competitive procurement process was the most important success factor.

Lastly it was found that all the interviewees from Australia and some of the ones from Hong Kong had their own organisation guidance or practice notes. This is highly recommended and especially useful for individuals and companies that are inexperienced with PPP practice.

A large number of differences were observed between the findings from the two jurisdictions. This result is logical as each jurisdiction will differ in practice, culture, geographical location, experience, tradition, and also politically, economically and socially. Hence it is of interest to compare these differences.

The private sector's perspective

Selecting respondents

The target private sector respondents of the interviews were practitioners of senior level and authority within the private sector with experience in PPP. A total of fourteen interviews were conducted, with seven interviews conducted in each jurisdiction. Interviewees of different backgrounds were purposely selected to compare the similarities and differences within the private sector. This was to make the findings even more representative if common findings are derived even though their backgrounds are so diverse.

Among the seven interviews conducted in Hong Kong, four of the interviewees work for local companies whereas the other interviewees each work for a Japanese, French and Australian company respectively. These companies comprise a property developer, three construction companies, an exhibition company, a law firm and a service provider.

Most of the Australian interviewees work for companies of their own country. Again the majority (four out of seven) of these are construction companies. The other three companies include a project management company, a bank and a credit rating company. Details of these experts are shown in Tables 4.10 and 4.11 for Hong Kong and Australian interviewees respectively.

Table 4.10 List of interviewees from the private sector in Hong Kong

No.	Position of interviewee	Organisation of interviewee	Experience of interviewee
PR1	Project Advisor	Local Property Developer	Strong interest in PPP, previously bid for several PPP projects but unsuccessful
PR2	Deputy General Manager	Japanese Construction Company working in Hong Kong	Involved with the design and construction of several PPP projects
PR3	General Manager	Local Exhibition Company	Previously a general manager for a social PPP project
PR4	Engineering Director	French Construction Company working in Hong Kong	Involved with the design and construction of several PPP projects
PR5	Executive Director	Local Construction Company	Conducted PPP projects in other countries
PR6	Managing Partner	Australian Law Firm working in Hong Kong	Acted as legal adviser for PPP projects in other countries
PR7	Commercial Manager	Local Service Provider	Involved with an ongoing PPP project

Source: Chan *et al.* 2008e (with permission from Chinese Research Institute of Construction Management).

Table 4.11 List of interviewees from the private sector in Australia

No.	Position of interviewee	Organisation of interviewee	Experience of interviewee
PR8	Executive Manager	Local Construction Company	Involved with the design and construction of many PPP projects
PR9	Manager	Local Construction Company	Involved with the design and construction of many PPP projects
PR10	Manager	Local Construction Company	Involved with the design and construction of many PPP projects
PR11	Manager	Local Construction Company	Involved with the design and construction of many PPP projects
PR12	Director	Local Project Management Company	Acted as the project director for many PPP projects
PR13	Head	Local Bank	One of the main financiers of PPP projects globally
PR14	Managing Director	American Credit Rating Company	Advised bidders on PPP projects

Interview findings

Table 4.12 shows a summary of the responses to each question given by the fourteen interviewees. The number of times that each response was given was tallied. Where the response was only given once it was judged insignificant for further analysis. The responses given more than once were tabulated and further analysed as shown in Tables 4.13 to 4.20. The numbers in brackets represent the number of times the response was mentioned by interviewees.

Local and international experience in conducting PPP projects

Table 4.13 presents the findings that were mentioned more than once by the interviewees for Question 1: 'Which PPP projects has your company been involved in?' The results showed that interviewees from both jurisdictions mentioned two responses more than once.

Four interviewees from Hong Kong and two from Australia mentioned that they had participated in both 'Local and international projects'. On the other hand, five interviewees in Australia and two in Hong Kong mentioned that they had been involved with 'Local projects' only.

The findings showed that the majority of Hong Kong interviewees had participated in projects both locally and abroad, whereas most of the Australian interviewees had participated in local projects only. Another observation is that none of the interviewees had participated in international projects only. Although Hong Kong has had a long history in BOT projects it has still conducted far fewer of these types of projects compared to Australia. Therefore most of the PPP expertise utilised in Hong Kong today has been sourced from overseas rather than trained locally. On the other hand, their extensive use of PPP in Australia has built up resources of their own to cater for these projects (Infranews 2008).

Implementation process of PPP projects

Thirteen aspects related to the implementation of PPP projects were mentioned by the interviewees. Only one of these was mentioned twice by the Australian interviewees: 'Reduce competition' (Table 4.14). The two Australian interviewees reported that there has been too much competition in the procurement process of PPP projects. They further suggested that the number of competitors involved in the process should be reduced. The transaction process for PPP projects can be both costly and lengthy. Hence the competitors need certainty that they have a high chance of winning the bid before they enter the process. If there are fewer competitors, they have more chance of success. Publications have also reported the problems related to the transaction process of PPP projects causing the private sector to become reluctant to continue participating in them (Ahadzi and Bowles 2004).

Table 4.12 Summary of responses from private sector interviewees

| | Hong Kong interviewees | | | | | | | | Australian interviewees | | | | | | | |
	PR1	PR2	PR3	PR4	PR5	PR6	PR7	Total	PR8	PR9	PR10	PR11	PR12	PR13	PR14	Total
1. Which PPP projects has your company been involved in?																
Local projects	✓		✓				✓	3	✓	✓	✓	✓	✓			5
Local and international projects		✓		✓	✓	✓		4						✓	✓	2
2. Please describe the implementation process in these projects.																
Increase competition	✓							1								0
Reduce competition								0	✓	✓						2
Refer to Efficiency Unit Guidelines		✓						1								0
Government and authority co-promoted			✓					1								0
Consultations before project start				✓				1								0
Not initiated by Treasury Department					✓			1								0
Pre-qualification exercise						✓		1								0
Government focus on administration							✓	1								0
Based on Victoria Partnerships projects								0			✓					1
Costly bidding process								0				✓				1
Banks take lead in project transaction								0						✓		1
Same as traditional								0							✓	1
Consist of specialists								0					✓			1

	Hong Kong interviewees								Australian interviewees							
	PR1	PR2	PR3	PR4	PR5	PR6	PR7	Total	PR8	PR9	PR10	PR11	PR12	PR13	PR14	Total
3. What were the major reasons for adopting PPP in these projects?																
Private sector expertise	✓	✓	✓		✓	✓		5			✓					1
Developing countries	✓							1								0
Private sector's cost and time certainty		✓						1			✓					1
Win–win situation				✓				1								0
Value for money				✓		✓	✓	3								0
Private sector efficiency					✓		✓	2			✓					1
Transfer of risks					✓			1		✓			✓			2
Large projects								0	✓							1
Government need	✓		✓					2				✓	✓	✓	✓	4
4. How do you think PPP compares with traditional procurement methods?																
Better integration	✓		✓					2								0
Private sector involvement		✓						1								0
Benefits for early completion		✓						1								0
Better value for money				✓		✓		2					✓			1
Performance based				✓	✓			2								0
Consider project life					✓			1								0
Differentiated product							✓	1			✓					1
Larger projects								0	✓				✓	✓		3

Continued overleaf

	Hong Kong interviewees								Australian interviewees							
	PR1	PR2	PR3	PR4	PR5	PR6	PR7	Total	PR8	PR9	PR10	PR11	PR12	PR13	PR14	Total
Increased efficiency and speed			✓			✓		2		✓			✓			2
More rigorous tendering process		✓						1		✓		✓				2
Paid according to schedule								0			✓					1
Different risk profiles								0			✓		✓	✓		3
Large emphasis on relationships								0			✓					1
No difference in terms of rating								0							✓	1
More complex								0					✓			1
More politically challenging								0					✓			1
5. Which type of project would your company be most interested in applying PPP?																
All	✓		✓	✓	✓		✓	5							✓	1
Any except transportation		✓						1								0
Projects with subsidy			✓					1								0
Projects with appropriate risk allocation					✓			1					✓			1
Projects with prospect of success						✓	✓	2		✓						1
Where land is not issue						✓		1								0
No staffing issues						✓		1								0
Recurrent expenditure already budgeted						✓		1								0
Projects with business case							✓	1						✓		1
Economic infrastructure								0	✓	✓				✓		3

	Hong Kong interviewees								Australian interviewees							
	PR1	PR2	PR3	PR4	PR5	PR6	PR7	Total	PR8	PR9	PR10	PR11	PR12	PR13	PR14	Total
Projects with clear objectives								0		✓						1
Social infrastructure								0			✓	✓	✓			3
6. What do you feel are the key performance indicators in a PPP project?																
Economics	✓		✓		✓	✓		4	✓		✓	✓	✓		✓	5
Risk	✓							1	✓	✓	✓					3
Control	✓							1								0
Refer to Efficiency Unit guide		✓						1								0
Contract terms (operating parameters)			✓				✓	2						✓		1
Traditional KPIs (time, cost and quality)				✓				1								0
Public sector comparator					✓			1								0
Service						✓		1								0
Time								0			✓	✓				2
7. In general, what do you think are the critical success factors leading to successful PPP projects?																
Economically viable	✓					✓		2	✓						✓	2
No Legislative Council interference		✓						1								0
Good relationships/partnering spirit			✓	✓				2		✓		✓				2
Clear project objectives/timeline				✓	✓	✓		3		✓				✓		2
Private party possess expertise				✓				1				✓				1
Appropriate allocation of risks				✓	✓			2	✓							1

Continued overleaf

	Hong Kong interviewees								Australian interviewees							
	PR1	PR2	PR3	PR4	PR5	PR6	PR7	Total	PR8	PR9	PR10	PR11	PR12	PR13	PR14	Total
Government support (champion)				✓		✓		2		✓						1
Transparent process						✓		1								0
Flexibility for innovation								0		✓						1
Robust contract								0			✓					1
Positive media								0			✓		✓			2
8. Does your company have any in-house guidance/practice notes on PPP implementation?																
Yes								0						✓		1
Other materials available		✓	✓	✓	✓			4	✓	✓	✓				✓	4
No	✓					✓	✓	3				✓	✓			2

Table 4.13 Question 1: 'Which PPP projects has your company been involved in?'

Hong Kong interviewees	Australian interviewees
Local and international projects (4)	Local projects (5)
Local projects (3)	Local and international projects (2)

Table 4.14 Question 2: 'Please describe the implementation process in these projects.'

Hong Kong interviewees	Australian interviewees
	Reduce competition (2)

Major reasons for adopting PPP projects

As shown in Table 4.15 four reasons were mentioned more than once by the Hong Kong interviewees and two by the Australians. Amongst these reasons one was mentioned by both groups: 'Government need'. This reason was mentioned the most at four times by the Australian interviewees, only twice by the Hong Kong interviewees. PPP in Australia were originally initiated because of financial shortages in the state governments to deliver public infrastructure and services (English and Guthrie 2003). The findings from the interviews have shown that many of the experts still regard this as an important reason to adopt PPP.

Mentioned the most, at five times, by Hong Kong interviewees was 'Private sector expertise'. The Hong Kong government has been able to enjoy a comfortable financial reserve. Hence other advantages such as the added benefits of the private sector's expertise and efficiency have been motivators to adopt PPP projects (Borzel and Risse 2005). Therefore 'Private sector efficiency' was also mentioned by the Hong Kong interviewees twice. Other reasons mentioned by the Hong Kong interviewees include 'Value for money', which has also been studied by previous researchers (Grimsey and Lewis 2005; Heald 2003).

In Australia 'Transfer of risks' was also mentioned twice by the interviewees. Risk is probably the most extensively studied aspect of the PPP method. Numerous studies have already been conducted on risk identification, allocation, treatment, and so on (Akbiyikli and Eaton 2004; Chan *et al.* 2008b; Hodge 2004; Jin and Doloi 2008; Li *et al.* 2004; Li *et al.* 2005b; Ng and Loosemore 2007; Shen and Wu 2005; Shen *et al.* 2006; Sun *et al.* 2008; Thomas *et al.* 2003; Thomas *et al.* 2006; Wibowo and Kochendörfer 2005)

Table 4.15 Question 3: 'What were the major reasons for adopting PPP in these projects?'

Hong Kong interviewees	Australian interviewees
Private sector expertise (5)	Government need (4)
Value for money (3)	Transfer of risks (2)
Private sector efficiency (2)	
Government need (2)	

Comparing PPP with traditional procurement methods

Four differences mentioned more than once between PPP and traditional procurement methods were suggested by each group of interviewees (Table 4.16). Only one of these was similar for both groups of interviewees: 'Increased efficiency and speed'. Efficiency and speed have been known as reasons for implementing public works projects by PPP (Borzel and Risse 2005; Bovaird 2004). The public sector tends to be more laid back compared to the private sector that tends to be more motivated due to commercial reasons (Sharma 2007).

Other differences mentioned more than once by the Hong Kong interviewees include 'Better integration', 'Better value for money' and 'Performance based'. Again these features have been highlighted by previous researchers. Nisar (2007) also agreed that a key aspect of PPP is the integration between partners. Many studies have also been conducted into the topic of value for money, as technically public projects are procured by the PPP model only when value for money can be achieved (Grimsey and Lewis 2005; Heald 2003; Nisar 2007). A key feature of PPP projects is that there is a heavy emphasis on project and private sector performance (Dhaene 2008).

The Australian interviewees also mentioned three other differences. Factors mentioned three times include 'Larger projects' and 'Different risk profiles'. Shen *et al.* (2006) mentioned that the best PPP projects should be ones that are mega-scale. As mentioned previously in this chapter, risks are an important part of PPP projects throughout the life cycle. The last difference mentioned twice was a 'More rigorous tendering process', which is another key feature of PPP projects (Ahadzi and Bowles 2004).

Table 4.16 Question 4: 'How do you think PPP compares with traditional procurement methods?'

Hong Kong interviewees	Australian interviewees
Better integration (2)	Larger projects (3)
Better value for money (2)	Different risk profiles (3)
Performance based (2)	More rigorous tendering process (2)
Increased efficiency and speed (2)	Increased efficiency and speed (2)

Projects of interest to the private sector

Table 4.17 shows the types of PPP projects that the interviewees would be most interested in. Five of the Hong Kong interviewees mentioned 'All'; two suggested 'Projects with prospect of success'. The private sector participates in PPP projects not because they have to but for commercial benefit, hence it is logical that they expect the project to be successful.

Three Australian interviewees mentioned that they would be interested in 'Economic infrastructure' and 'Social infrastructure'. Public works projects normally comprise either economic or social infrastructure works. Economic projects are normally those where the income is collected directly from the end-user, for example toll roads, railways, and so on. On the other hand, social infra-structure projects are normally supported by a regular fee paid by the government, for example schools, hospitals. Both of these were mentioned by the Australian interviewees indicating that most types of public works projects are supported by the private sector.

Similarly the findings showed that the Hong Kong interviewees were interested in all PPP projects. The main difference is probably that Hong Kong's experience in PPP projects has mainly been in economic infrastructure (Mak and Mo 2005), hence they did not further break public works projects up into categories. Australia has conducted a larger range of different-natured PPP projects hence they tend to split them into categories.

Key performance indicators of PPP projects

Table 4.18 shows the key performance indicators mentioned by the interviewees. The top key performance indicator identified by both groups of interviewees was 'Economics'. This key performance indicator was mentioned four times by the Hong Kong interviewees and five times by the Australian interviewees. Money has often been used as a measure to quantify performance, especially in the private sector where motives for participating in PPP projects are often related to commercial aspects (Sharma 2007).

Other key performance indicators mentioned by the Hong Kong interviewees include 'Contract terms (operating parameters)' which was mentioned twice. Contracts can be used to measure the outputs of a project (Entwistle and Martin 2005).

Table 4.17 Question 5: 'Which type of project would your company be most interested in applying PPP?'

Hong Kong interviewees	Australian interviewees
All (5)	Economic infrastructure (3)
Projects with prospect of success (2)	Social infrastructure (3)

The Australian interviewees mentioned 'Risk' three times and 'Time' twice. As mentioned before, risk is an important aspect of the PPP arrangement. On the other hand time is also a key performance indicator for traditional projects and highly related to the commercial aspects of the project.

Critical success factors leading to successful PPP projects

Table 4.19 shows that the Hong Kong interviewees mentioned five critical success factors more than once, whereas the Australian interviewees mentioned four. Three of these critical success factors were similar, including 'Clear project objectives/timeline' mentioned three times by the Hong Kong interviewees and twice by the Australian interviewees, and 'Economically viable' and 'Good relationships/partnering spirit' mentioned twice by both groups of interviewees. The importance of clear project objectives and timeline has also been incorporated into governmental guidance notes (Efficiency Unit 2003). In Zhang's (2005a) study he looked at several groups of potential critical success factors; one of these was identified as economic viability. Partnering spirit between the parties is a vital aspect of PPP projects (Gunnigan and Eaton 2006).

Other critical success factors also mentioned twice by the Hong Kong interviewees include 'Appropriate allocation of risks' and 'Government support (champion)'. Again, risks are mentioned as being highly important in the PPP arrangement. In addition, Zhang (2005a) derived from his study that government support is the second most significant critical success factor under the group of 'Favourable investment environment'.

The Australian interviewees also mentioned 'Positive media' twice. The effect of positive media is incredibly important for the success of a PPP project. In the case of the Cross City Tunnel in Sydney, Australia, the negative portrayal of the media caused the effects of its faults to intensify (Chan *et al.* 2008b).

In-house guidance/practice notes

Table 4.20 shows that both groups of interviewees had mentioned 'Other materials available' instead of in-house guidelines/practice notes for PPP implementation. This finding is probably because the private sector tends to follow the guidelines of their client in the public sector, hence there is no need for them to derive their

Table 4.18 Question 6: 'What do you feel are the key performance indicators in a PPP project?'

Hong Kong interviewees	Australian interviewees
Economics (4)	Economics (5)
Contract terms (operating parameters) (2)	Risk (3)
	Time (2)

Table 4.19   Question 7: 'In general, what do you think are the critical success factors leading to successful PPP projects?'

Hong Kong interviewees	Australian interviewees
Clear project objectives/timeline (3)	Clear project objectives/timeline (2)
Economically viable (2)	Economically viable (2)
Good relationships/partnering spirit (2)	Good relationships/partnering spirit (2)
Appropriate allocation of risks (2)	Positive media (2)
Government support (champion) (2)	

own guidance materials. Governmental guidance notes are readily available for the general public's use (Efficiency Unit 2012c; Partnerships Victoria 2008b). Nevertheless it was found that the organisations of most of the interviewees had conducted some kind of research on the topic of PPP to broaden their knowledge of the process. Some of the interviewees mentioned 'No' to this final question (three in Hong Kong and two in Australia) showing the lack of need for in-house guidance/practice notes.

Summary of findings from the private sector

This section has studied the private sector's perspective on procuring public works projects, via findings from fourteen interviews conducted in Hong Kong and Australia. Experts with experience in PPP projects and research were invited to answer eight questions related to PPP implementation. The results found that interviewees from Hong Kong had participated in projects both locally and internationally, whereas the Australian interviewees had participated mainly in local projects. Australia has conducted many more PPP projects compared to Hong Kong, so they have built up and trained their own resources and expertise over time. Hong Kong, on the other hand, may not have the necessary talents, so it is important for them to either gain overseas experience or import their expertise. With the increasing number of PPP projects conducted in Hong Kong there is a need to start training their own people.

Table 4.20   Question 8: 'Does your company have any in-house guidance/practice notes on PPP implementation?'

Hong Kong interviewees	Australian interviewees
Other materials available (4)	Other materials available (4)
No (3)	No (2)

The implementation process of PPP projects can vary depending on the project itself, hence a large range of descriptions were given by the interviewees. Only 'Reduce competition' was repeatedly mentioned. Reducing competition in the tendering process would mean that the chance of winning would be greater for the bidding party. Hence in future projects the government should consider a suitable number of competitors in the tendering process to avoid the private sector's loss in terms of time, money and especially lack of confidence in conducting future PPP projects.

Only one reason for adopting PPP projects was mentioned by both groups of interviewees which was 'Government need'. Other reasons mentioned by the Hong Kong respondents included private sector expertise, value for money and private sector efficiency. All these reasons have been mentioned in many literature pieces for being typical features and attractions of the PPP model, hence the findings have further verified previous studies.

Another major reason for adopting PPP is for risk transfer, as mentioned by the Australian interviewees; this topic is probably the most discussed. It is likely that the Hong Kong interviewees are familiar with delivering public works projects themselves, and are therefore comfortable with the practice. They did not highlight risk transfer as a reason for adopting PPP projects as they are less aware of the risks. But according to previous cases around the world, the Hong Kong government should consider an appropriate allocation of risks in order for public projects to be delivered more effectively.

The common difference between PPP and traditional methods highlighted by both groups of interviewees was increased efficiency and speed in PPP projects. Added value has become the foremost reason to apply PPP projects these days. Governments have realised that the private sector tends to be more motivated and driven, for commercial reasons. Other differences highlighted by the Hong Kong interviewees include: better integration; better value for money; and being performance based. Those mentioned by the Australian interviewees include: larger projects; different risk profiles; and a more rigorous tendering process.

The findings also showed that the interviewees were interested in all sorts of projects. In Hong Kong the interviewees did not further categorise PPP projects, probably because there have been few PPP projects in Hong Kong, which have tended to be transportation related, and so the variety has not been large. In Australia, on the other hand, there have been various types of PPP projects conducted; hence the interviewees categorised them into social or economic types of projects. The preference for conducting these two groups of projects was similar.

Both groups of interviewees agreed that economic performance indicators were the most appropriate. The Hong Kong interviewees also recommended that the contract terms could be used to measure performance. The Australian interviewees also mentioned risk and time as key performance indicators. Therefore the key performance indicators for PPP projects should focus heavily on the economics.

Five critical success factors were mentioned more than once by Hong Kong interviewees and four by the Australian interviewees. Amongst these, three were mentioned by both groups of interviewees: clear project objectives/timeline;

economically viable; and good relationship/partnering spirit. Clear objectives and timeline allow the private sector to keep on track over a long project period. Typically in PPP projects, the consortium's responsibility is over thirty years instead of a normal design and construct contract which will maybe span over five years. Hence clear objectives and timeline are particularly important. These should be well defined before the project commences. An economically viable project is also vital as the private sector is driven by commercial motives. In order to sustain such a long-term relationship over the project period, good relations/partnering spirit are also vital. The parties involved could consider utilising tools such as partnering workshops to increase their communication and understanding.

Other critical success factors mentioned by Hong Kong interviewees include appropriate allocation of risks and government support (champion). The Australian interviewees also mentioned positive media attention as a critical success factor. Finally, it was realised that the interviewees all had access to some kind of PPP guidance materials.

The researcher's perspective

Selecting respondents

The target respondents of the interviews were researchers with experience in PPP who neither belonged to nor acted for the public or private sectors. A total of seven interviews were conducted, with three in Hong Kong and four in Australia.

Amongst the three interviewees from Hong Kong, two were members of the Legislative Council in Hong Kong (one with a law background and the other with an engineering background). The third interviewee was an academic and researcher in PPP from a local university. Similarly, the Australian interviewees were all active researchers of the PPP topic from local universities. Due to the limited number of PPP projects conducted in Hong Kong (not including BOT type), fewer academics are involved with PPP-related research, hence two legislative councillors were selected, both having been known to publicly mention their interests in PPP. As their role tends to represent the general public rather than the public or private sector, it was believed that their position would be similar to the academics interviewed. Background details of these experts are shown in Table 4.21.

Interview findings

Table 4.22 shows a summary of the responses to each question given by the seven interviewees. The number of times that each response was given was tallied. Where the response was only given once it was believed to be insignificant for further analysis. For the responses given more than once, these were further analysed.

Table 4.21 List of researcher interviewees

No.	Jurisdiction	Position of interviewee	Organisation of interviewee	Experience of interviewee
R1	Hong Kong	Member of Legislative Council (legal background)	Legislative Council of the Hong Kong Government	Commissioned a working group to analyse the feasibility of PPP
R2	Hong Kong	Member of Legislative Council (engineering background)	Legislative Council of the Hong Kong Government	Supporter of PPP and included in his campaign to push the development
R3	Hong Kong	Professor	Local University	Conducted his own PhD in the area of PPP and active researcher
R4	Australia	Professor	Local University	Active researcher
R5	Australia	Professor	Local University	Involved with producing governmental guidelines, training, courses and research for PPP
R6	Australia	Professor	Local University	Active researcher
R7	Australia	Professor	Local University	Involved with producing governmental guidelines, training, courses and research for PPP

Source: Cheung *et al.* 2010b (with permission from Emerald Group Publishing Ltd).

Research on local case studies

The first question that the interviewees were asked to answer was 'Have you conducted any research looking at local case studies?' All interviewees responded that they had conducted PPP case studies and research both locally and overseas. In general, it can be summarised that the interviewees are active experienced researchers in the field of PPP.

Comparing PPP with traditional procurement methods

The interviewees were further asked 'How would you compare PPP with traditional procurement methods?' Thirteen different responses were given, but only four of these were mentioned more than once. Responses which were each mentioned twice included: 'PPP is a partnership arrangement'; 'PPP projects have high tendering/transaction costs'; 'Different risk profiles'; and 'Private sector more innovative/efficient'.

Mentioned by the Hong Kong interviewees only was 'PPP is a partnership arrangement' and 'PPP projects have high tendering/transaction costs'. The Efficiency Unit of the Hong Kong government has been actively involved in pushing the

Table 4.22 Summary of interview findings with researchers from Hong Kong and Australia

	Hong Kong interviewees			Australian interviewees				
	R1	R2	R3	R4	R5	R6	R7	Total
1. Have you conducted any research looking at local case studies?								
Yes	✓	✓	✓	✓	✓	✓	✓	7
2. How would you compare PPP with traditional procurement methods?								
Clear project objectives	✓							1
PPP utilises public resources		✓						1
PPP is a partnership arrangement		✓	✓					2
PPP projects have high tendering/ transaction costs		✓	✓					2
PPP projects tend to be completed on time			✓					1
Income of PPP projects can be dependent on market			✓					1
Construction costs of PPP projects are more expensive			✓					1
PPP considers maintenance			✓					1
Different risk profiles			✓		✓			2
More expensive for private sector to borrow money			✓					1
Private sector more innovative/efficient			✓	✓				2
PPP focuses on service delivery					✓			1
PPP improves public procurement							✓	1
3. Which type of project do you feel is best suited to use PPP?								
Project which government lacks funding	✓		✓					2
Project dependent	✓	✓						2
Projects with few competitors			✓					1
Large projects				✓	✓			2
Expensive projects				✓				1
Quantifiable income stream				✓				1
Scope for innovation					✓			1
Toll ways							✓	1
4. What do you feel are the key performance indicators in a PPP project?								
Profits	✓							1
Project dependent			✓		✓			2
Traditional KPIs: quality, time and cost			✓					1
Should be defined by private sector				✓				1

Continued overleaf

	Hong Kong interviewees			Australian interviewees				
	R1	R2	R3	R4	R5	R6	R7	Total
Service outcomes					✓			1
Contract compliance							✓	1
Proactive managers							✓	1
5. In general, what do you think are the critical success factors leading to successful PPP projects?								
Clear project objectives	✓	✓						2
Timeline and milestones foreseeable	✓							1
Transparent process		✓					✓	2
Public consultation		✓						1
Project dependent		✓			✓			2
Clear legal structure and regulation mechanism			✓					1
Market need			✓				✓	2
Technical and financial capability of concessionaire			✓					1
Champion with authority					✓			1
Roles clearly defined and related to each other							✓	1
Need to budget money for project amount							✓	1
Right timing							✓	1
Strong and robust contract							✓	1
Commitment of partners							✓	1

Source: Cheung *et al.* 2010b (with permission from Emerald Group Publishing Ltd).

movement of PPP in Hong Kong. In one of their latest guidelines they mention the importance of the partnership arrangement (Efficiency Unit 2012a).

A common feature which is found in PPP projects is the high costs of tendering and transaction (Zhang 2005b). Hughes *et al.* (2001) reported that the costs associated with tendering are seen by the construction industry in the UK to be significant, typically quoted as ½–1 per cent of turnover; and 2–3 per cent of bid price for PPP bids. Furthermore, results from their study showed that building services contractors had calculated that up to 15 per cent of their turnover could be accounted for by 'unnecessary' tendering processes.

For the other two responses which were mentioned more than once ('Different risk profiles' and 'Private sector more innovative/efficient'), these were mentioned by interviewees from both jurisdictions. As mentioned previously in this section, one of the main reasons for implementing public projects by PPP is also for risk transfer. The National Stadium for the Beijing 2008 Olympic Games in China is an example of how key risk factors were appropriately passed to the private sector via the PPP model (Liu *et al.* 2009). Without doubt this project had been highly profiled, so the pressure to perform well increased the associated risks. The four most critical risks of this project included: (1) the irrational construction schedule for a project of this size and complexity; (2) possible cost overruns due to inexperience in delivering similar projects in China; (3) the small and limited market for large-scale sporting events in China; and (4) the lack of operational experience in similar previous projects in China. These factors were all related to the management, design, construction and operation of the project, which are also aspects that are considered to be best handled by the private sector; whereas the public sector's expertise lies in the area of public administration.

Another major difference observed between traditionally procured projects and PPP projects is the added innovation and efficiency of the private sector in PPP projects. The private sector in general tends to be more motivated because of financial drive, whereas the public sector parties are experts in policy making rather than innovation and efficiency. Studies have shown that by adopting PPP in public works projects, innovation and efficiency are achieved because of the private sector's contribution (Leiringer 2006)

Projects best suited to use PPP

The interviewees were asked 'Which type of project do you feel is best suited to use PPP?' in Question 3. Three out of the eight responses were mentioned twice by the interviewees; these included 'Project which government lacks funding' and 'Project dependent' which were mentioned by the Hong Kong interviewees and 'Large projects' mentioned by the Australian interviewees. In many jurisdictions which first started to adopt PPP, private financing was a major incentive for governments to adopt PPP, such as the UK and the state of Victoria in Australia. Therefore there has been a common association that PPP projects are about financing.

An example showing that Hong Kong does not need private sector financing is the recent Hong Kong Zhuhai Macau bridge project, where the governments of these three jurisdictions agreed to undertake the project costs without private sector input. The Hong Kong government alone has agreed to cover approximately 50 per cent of the costs, approximately HK$15.3 billion (approximately US$1.97 billion on 11 December 2012, Yahoo! Finance 2012) (South China Morning Post 2008).

The Hong Kong interviewees also mentioned that the suitability criteria of projects to be procured by PPP could be unique, depending on the project. The Australian interviewees mentioned that large projects would be suitable for the PPP model. Similarly, Price (2002) suggested that for some types of projects, especially those that are large or complex, a joint venture between the public and private sectors may prove advantageous.

Key performance indicators in PPP projects

Only one response was mentioned more than once, at twice by the interviewees for Question 4: 'What do you feel are the key performance indicators in a PPP project?' This was 'Project dependent' which was mentioned by interviewees from both jurisdictions. Six other responses were given by the interviewees for this question.

Critical success factors leading to successful PPP projects

For the final question, interviewees were asked: 'In general, what do you think are the critical success factors leading to successful PPP projects?' This question received the most responses, probably indicating that there are many critical success factors that could lead to successful PPP projects. But amongst these responses only four were mentioned more than once by the interviewees; these included 'Clear project objectives' which was mentioned by the Hong Kong interviewees only and 'Transparent process', 'Project dependent' and 'Market need' which were all mentioned by interviewees from both jurisdictions.

Zhang (2006) mentioned in his study that the public client often does not have clear objectives and priorities in infrastructure development through PPP. This often impairs the project development process. The client should clearly define its objectives and establish their relative importance and make sure the private sector shares these objectives. The probability of successful project delivery increases dramatically when both sectors have a common vision of the project to be developed. The *Partnerships Victoria Policy* (2000) mentions that where there is private sector involvement in major public infrastructure projects, the choice of contractors should be through a rigorous and transparent system of public tendering.

Similar to the responses for Questions 3 and 4, the interviewees also mentioned that the critical success factors for PPP projects would be dependent on the project because of their uniqueness. Lastly, Partnerships Victoria (2001) also mentioned that a key characteristic of Partnership Victoria projects (that is, PPP projects conducted under the Partnership Victoria's supervision) is market appetite. This implies that the project creates a genuine business opportunity which is likely to attract a sufficient number of private parties and create an effective and competitive bidding process.

Summary of findings from researchers

This section has presented the findings of seven interviews conducted with experienced researchers in the field of PPP from Hong Kong and Australia. The interviewees were asked to answer five questions related to the implementation of PPP. It was found that both groups of interviewees had conducted case studies and research in the field of PPP locally and internationally.

When considering the differences between traditionally procured projects and PPP projects, both groups of interviewees agreed that 'Different risk profiles' and 'Private sector more innovative/efficient' were the main differences. Other major differences between the two approaches mentioned by the Hong Kong interviewees included 'PPP is a partnership arrangement' and 'PPP projects have high tendering/transaction costs'. The types of project best suited to use PPP were not the same according to the two groups of researchers.

The Hong Kong interviewees recommended that 'Projects which government lacks funding' and 'Project dependent' are suitable criteria for PPP projects, whereas the Australian researchers believed that 'Large projects' would be more ideal.

Amongst the key performance indicators highlighted by the interviewees, 'Project dependent' was the only response given by both groups of interviewees.

From the large number of critical success factors suggested, 'Transparent process', 'Project dependent' and 'Market need' were the common ones highlighted by both groups of interviewees. Hong Kong interviewees also believed that 'Clear project objectives' would be an important critical success factor.

Chapter summary

This chapter has summarised the views of experienced practitioners from the public sector, private sector and renowned researchers from Hong Kong and Australia.

It was found that the public sector and researcher interviewees from both jurisdictions had conducted research in PPP and looked at international cases. From their experiences other governments should also adopt a similar practice. Although both jurisdictions had conducted research in PPP, the Australian private sector had conducted many more PPP projects than those in Hong Kong.

Many of the private sector interviewees believed that competition in the tendering process should be reduced. Hence governments should consider what the suitable number of bidders for a project should be.

The public sector and researcher interviewees agreed that the main difference between PPP and traditional projects is that in a PPP project there is the added advantage of the private sector's efficiency, expertise and management skills involved. The researcher interviewees also added that PPP projects and traditional projects differ in their risk profiles. On the other hand, the private sector highlighted that the main difference is that PPP can increase the efficiency and speed of projects.

The public sector interviewees agreed that projects which are economically viable are those suitable to be procured by PPP. The private sector, on the other hand, believed that projects are normally procured by PPP due to government need, but all types of projects should be suitable.

The public sector interviewees believed that the contractor's performance would be the key performance indicator for PPP projects. The private sector interviewees believed that economic performance indicators are the most important. On the other hand, the researcher interviewees believed that each project is unique so the key performance indicators would be as well.

The project objectives being well defined and a partnering spirit/commitment/ trust were the critical success factors mentioned by both the public and private sector interviewees. The private sector interviewees also added that the project must be economically viable. The researcher interviewees highlighted some further critical success factors including projects having a transparent process and there being a market need (this factor is very closely linked with the project being economically viable as mentioned by the private sector interviewees).

Most of the public and private sector interviewees had their own organisation guidance/practice notes. This practice was highly recommended.

The views presented in this chapter are believed to be of interest to all practitioners involved with PPP projects. The findings also form a comparison between the views of PPP practitioners in Hong Kong and Australia, and draw similarities irrespective of the differences in jurisdictions.

5 Attractive and negative factors of procuring public works by public–private partnership

Introduction

This chapter will specifically consider whether public–private partnership (PPP) should be used to procure public works projects by studying the attractive and negative factors for adopting PPP. Cheung *et al.* (2010c) and Cheung and Chan (2009b) conducted a questionnaire survey with industrial practitioners in Hong Kong and Australia, and compared it with findings obtained by Li (2003) in the United Kingdom. The respondents were requested to rank the importance of fifteen attractive factors and thirteen negative factors for adopting PPP.

Collection of research data

An empirical questionnaire survey was undertaken in Hong Kong and Australia to analyse the attractive and negative factors of adopting PPP. The target survey respondents of the questionnaire included all industrial practitioners from the public, private and other sectors. These respondents were requested to rate their degree of agreement against each of the identified attractive and negative factors according to a five-point Likert scale (1 = least important and 5 = most important).

Target respondents were selected based on two criteria: (1) they must possess adequate knowledge in the area of PPP; and (2) they must have hands-on experience with PPP projects, experience in conducting PPP research or have followed very closely the development of PPP. Survey questionnaires were sent to ninety-five target respondents in Hong Kong and eighty target respondents in Australia. It was anticipated that some of these target respondents would have colleagues and personal connections knowledgeable in the area of PPP to participate in this research study as well; hence some of the respondents were dispatched five blank copies of the survey form. A total of thirty-five completed questionnaires from Hong Kong and eleven from Australia were returned, representing response rates of 36 per cent and 9 per cent, respectively. The lower response rate in Australia was expected as the questionnaire was administered from Hong Kong, hence geographical complications were perceived.

It must be noted that the number of responses in the Kendall's concordance analysis may not always be thirty-four for Hong Kong and eleven for Australia, as these respondents may not have ranked all the factors. Therefore in some cases not all responses may have been suitable for subsequent statistical analyses.

The questionnaire respondents comprised experienced practitioners from the industry. As shown in Figures 5.1 and 5.2 approximately half of the respondents in Hong Kong and Australia possessed twenty-one years or more of industrial experience. Figures 5.3 and 5.4 provide a breakdown of questionnaire respondents who have been involved with PPP projects.

Given the few Build Operate Transfer (BOT)/PPP projects conducted in Hong Kong, it was a surprise to find that 33 per cent of the respondents had gained previous experience. Without doubt some of these may have had experience with local BOT projects or PPP projects overseas, but still the experience of these respondents confirmed the quality of the responses from the survey conducted. In addition, amongst those respondents who have acquired experience with PPP projects, 10 per cent had previously been involved with at least five projects.

In Australia, many more PPP projects have been conducted so it was unsurprising to find that approximately 90 per cent of the respondents have participated in PPP projects before, with two-thirds of these respondents having participated in at least five PPP projects. Once again this ensures the value and reliability of the findings.

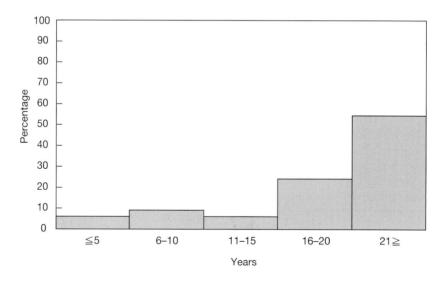

Figure 5.1 Histogram showing the number of years of working experience in construction industry for the Hong Kong survey respondents (Cheung and Chan 2011a; Cheung *et al.* 2012) (with permission from Emerald Group Publishing Ltd and the American Society of Civil Engineers)

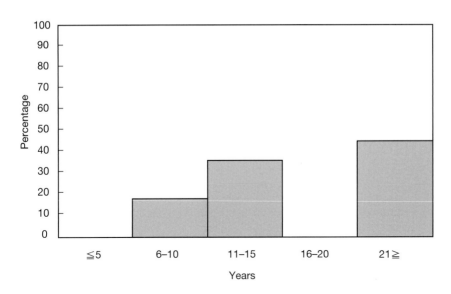

Figure 5.2 Histogram showing the number of years of working experience in construction industry for the Australian survey respondents (Cheung *et al.* 2012) (with permission from Emerald Group Publishing Ltd)

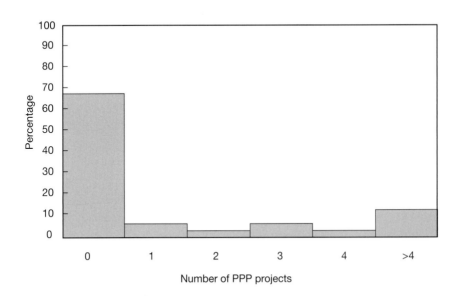

Figure 5.3 Histogram showing the number of PPP projects the Hong Kong survey respondents have been involved with (Cheung and Chan 2011a; Cheung *et al.* 2012) (with permission from Emerald Group Publishing Ltd and the American Society of Civil Engineers)

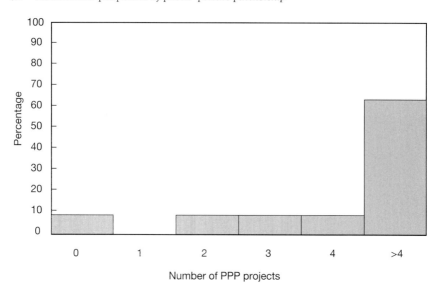

Figure 5.4   Histogram showing the number of PPP projects the Australian survey respondents have been involved with (Cheung *et al.* 2012) (with permission from Emerald Group Publishing Ltd)

Survey results

The attractive and negative factors for adopting PPP were assessed from different perspectives of the respondent groups from Hong Kong, Australia and the UK (results obtained by Li (2003) from his survey). The means for each administrative system were calculated and ranked in descending order of importance as shown in Tables 5.1 and 5.2. A large number of differences were observed between the findings of the three jurisdictions. This result is logical as each jurisdiction will differ in practice, culture, geographical location, experience and tradition, as well as politically, economically and socially. Hence it is of interest to compare these differences.

Ranking of attractive factors for adopting PPP

Fifteen attractive factors for adopting PPP were rated by the respondents (Table 5.1). The findings showed that the top three attractive factors ranked in Hong Kong were:

1   Provide an integrated solution (for public infrastructure and/or services);
2   Facilitate creative and innovative approaches;
3   Solve the problem of public sector budget restraint.

The first and second attractive factors ranked by Hong Kong respondents were also ranked identically by respondents in Australia. Ranked third in Australia was

'Save time in delivering the project'. The results show that both Hong Kong and Australia ranked efficiency-related attractive factors more importantly. Although financial drive in general is a major reason for adopting PPP, these respondents did not rank it as the top attractive factor.

Since Hong Kong has enjoyed abundant financial reserves in hand and budget surplus over the past few years, these have allowed Hong Kong to pay for their public works projects upfront. The government officials generally did not see the need to borrow money when they could provide the cash more cheaply. Hence efficiency was a more important attractive factor that could really induce Hong Kong to adopt PPP.

Similarly in Australia, although financial reasons may previously have been the motive for adopting PPP this is no longer the case. The state of Victoria in Australia first utilised PPP in order to deliver PPP projects using private sector money, but as the financial situation has improved and more experience has been gained, the Australians have realised other benefits of adopting PPP rather than for financial reasons alone.

On the contrary, in the UK economic-related factors were ranked much higher in 2003. The top attractive factor ranked in the UK was 'Transfer risk to the private partner'. Ranked second was the same attractive factor that came third in Hong Kong. And ranked third was 'Non-recourse or limited recourse to public funding'. Similar to Victoria, the UK also initially introduced PPP for financial reasons. But the situation today may also no longer be the same. The results from Li's (2003) study were obtained many years before those presented in this study, hence certain changes in the adoption of PPP and the attitude of practitioners are anticipated. Therefore as mentioned before the results in Li's (2003) study can only be used as a reference.

The first attractive factor ranked in Hong Kong 'Provide an integrated solution (for public infrastructure/services)' was also positioned first in the ranking for Australia. The rankings have demonstrated that this attractive factor was regarded as equally important to them irrespective of location.

PPP is an integrated solution in that a private consortium is responsible for all the functions of design, building, financing, operation and maintenance. This bundling can allow the partners to take advantage of a number of efficiencies and increase economies of scale and scope (European Commission Directorate 2003). For instance, the contractor's detailed knowledge of the project design and the materials utilised allows it to develop a tailored maintenance plan over the project life that anticipates and addresses needs as they occur, thereby reducing the risk that issues will go unnoticed or unattended and then deteriorate into much more costly problems. In the UK, this factor was ranked eighth showing only a medium level of importance.

The second attractive factor ranked by respondents from Hong Kong and Australia was 'Facilitate creative and innovative approaches'. In the UK, Li (2003) found that this attractive factor was rated seventh amongst fifteen attractive factors for PPP. This observation manifests that Hong Kong and Australia have a much greater urge for creativity and innovation in PPP projects compared to the UK. In the UK there has been a tendency for the local government to deliver packages of projects which are very similar, such as for schools. The creativity difference between these projects is often minimal. Nevertheless, practitioners in Hong Kong have publicly expressed the need and importance for creativity and innovation in PPP projects (Kwan 2005; Ho 2005).

The third attractive factor rated by respondents from Hong Kong, 'Solve the problem of public sector budget restraint', was also positioned highly at second place in the ranking of respondents from the UK. Therefore, both administrative systems perceived this attractive factor as highly important for launching PPP projects.

The financing of public sector projects has been recognised as one of the key initial driving forces for implementing PPP schemes internationally. Many experienced practitioners of PPP believe that PPP involves many other attractions besides financing, and that financial motivations should not be taken as the sole reason for adopting PPP. However, financial reasons are frequently the initial attractive factors for administrative systems adopting PPP. This financial attractive factor is undoubtedly very appealing for governments across the world, especially when public money is to be spent amongst competing needs. Therefore, it is not surprising that both groups of respondents have rated this attractive factor highly, but with a subtle difference in emphasis.

Contrastingly, the Australian respondents ranked this factor thirteenth amongst the fifteen attractive factors. This could imply that Australia currently does not face any major restraints in the public sector budget. But the views of more respondents from Australia should be sought before confirming this conclusion, as there were only eleven respondents from the survey conducted in Australia.

The mean values for the attractive factors as rated by Hong Kong respondents ranged from 2.94 to 3.79. This observation has reflected that the variation in their responses are relatively small: only 0.85 for Hong Kong. In Australia and the UK the means ranged from 2.36 to 4.45 and 1.82 to 3.98 respectively. The corresponding differences in means were 2.09 and 2.16 respectively. The differences in means were shown to be much higher for the survey conducted in Australia and the UK compared to Hong Kong. This finding shows that the Hong Kong respondents rated the fifteen attractive factors much more consistently, whereas in Australia and the UK the respondents showed a much larger variation.

As the respondents were asked to rate the fifteen attractive factors according to a Likert scale from 1 to 5 (1 = least important and 5 = most important), a value above 3 would represent that the attractive factor is of importance. Amongst the attractive factors only one was ranked below 3 in the Hong Kong rank. This attractive factor was 'Technology transfer to local enterprise' which scored 2.94 and was also ranked bottom in Hong Kong. This is probably because the immediate results of this attractive factor could not be seen and therefore the other fourteen attractive factors were relatively more favourable. In Australia and the UK (Li 2003) this attractive factor was rated 3.18 and 1.82 respectively, showing that the first set of respondents disagreed but the latter set of respondents agreed with those from Hong Kong. The other fourteen attractive factors in the Hong Kong rank were rated a score between 3 and 4.

Table 5.1 Mean scores and rankings for the attractive factors of PPP

Attractive factors	Hong Kong			Australia			United Kingdom (Li 2003)		
	N	Mean	Rank	N	Mean	Rank	N	Mean	Rank
a. Solve the problem of public sector budget restraint	34	3.65	3	11	2.73	13	61	3.86	2
b. Provide an integrated solution (for public infrastructure/services)	33	3.79	1	11	4.45	1	61	3.05	8
c. Reduce public money tied up in capital investment	33	3.48	6	11	2.36	15	61	3.58	4
d. Cap the final service costs	34	3.26	10	11	3.55	6	61	3.56	5
e. Facilitate creative and innovative approaches	34	3.74	2	11	4.36	2	61	3.36	7
f. Reduce the total project cost	33	3.09	14	11	3.45	7	61	2.97	10
g. Save time in delivering the project	34	3.21	13	11	4.18	3	61	2.75	12
h. Transfer risk to the private partner	34	3.65	4	11	3.36	9	61	3.98	1
i. Reduce public sector administration costs	33	3.39	8	11	2.82	12	61	2.53	14
j. Benefit to local economic development	34	3.56	5	11	3.18	11	61	2.62	13
k. Improve buildability	33	3.24	11	11	3.73	5	61	3.03	9
l. Improve maintainability	34	3.32	9	11	4.18	4	61	3.36	6
m. Technology transfer to local enterprise	34	2.94	15	11	3.18	10	61	1.82	15
n. Non-recourse or limited recourse to public funding	34	3.21	12	11	2.64	14	61	3.61	3
o. Accelerate project development	34	3.47	7	11	3.36	8	61	2.95	11

Source: Cheung and Chan 2009b; 2011a; Cheung et al. 2010c (with permission from Professor George Zillante, Editor of Proceedings of the 34th Australasian Universities Building Education Association Annual Conference, Adelaide, the American Society of Civil Engineers and Emerald Group Publishing Ltd).

N = Number of survey respondents

In addition, on top of those factors the respondents were asked to rate, they were also given the opportunity to add others which would be of importance, but they did not do so.

Ranking of negative factors for adopting PPP

Thirteen negative factors for adopting PPP were rated by the survey respondents (Table 5.2). The top three negative factors ranked by Hong Kong respondents included:

1   Lengthy delays because of political debate;
2   Lengthy delays in negotiation;
3   Very few schemes have actually reached the contract stage (aborted before contract).

In Hong Kong, public works projects are often delayed and complicated by the need for public consultation; hence it is not surprising that 'Lengthy delays because of political debate' was the highest negative factor ranked by the Hong Kong respondents. This problem is well known for causing projects to be held back. For example, the West Kowloon Cultural District project has been cited as a typical example in Hong Kong where political interference has caused the project to be on hold for many years (Chan *et al.* 2007b). Initially there was much political debate within the Legislative Council as to whether this project should proceed as a PPP, especially whether the whole project with an estimated cost of US$25 billion (So 2009) should be handled by one single consortium instead of half a dozen consortia each sharing the pie. The local government was also alleged to be unclear about the long-term policy and objectives for this cultural development project, causing much criticism from the general public. In Australia and the UK (Li 2003) this factor was ranked of mediocre importance only at seventh and sixth position respectively, showing that they do not face a similar problem to Hong Kong.

Ranked second by respondents in Hong Kong and the UK (Li 2003) was 'Lengthy delays in negotiation'. Australia also ranked this factor relatively high at fourth place. This finding has shown that 'Lengthy delays in negotiation' are typical for PPP projects irrespective of geographical locations. Due to the size and complexity of PPP projects the procurement process has been known to be lengthy. This can be said to be a typical feature of PPP projects, therefore only projects that are of appropriate value and worthiness should consider PPP.

The third negative factor as ranked by Hong Kong respondents was 'Very few schemes have actually reached the contract stage (aborted before contract)'. The high ranking of this factor coincides with the previous argument about political debate in Hong Kong. As a result some projects had to be aborted because of political disagreement. This negative factor was ranked last in Australia and Li's (2003) survey. The experience of the Australians and the British in conducting PPP projects is much more successful, in that the respondents did not believe that few schemes would reach the contract stage. Without doubt they are much more experienced and hence more confident in launching PPP projects.

Table 5.2 Mean scores and rankings for the negative factors of PPP

Negative Factors	Hong Kong			Australia			The United Kingdom (Li, 2003)		
	N	Mean	Rank	N	Mean	Rank	N	Mean	Rank
a. Reduce the project accountability	34	2.79	12	11	2.00	11	61	1.90	11
b. High risk relying on private sector	34	3.09	10	11	2.27	8	61	2.22	10
c. Very few schemes have actually reached the contract stage (aborted before contract)	34	3.41	3	11	1.36	13	61	1.71	13
d. Lengthy delays because of political debate	34	3.82	1	11	2.55	7	61	2.48	6
e. Higher charge to the direct users	34	3.26	9	11	2.18	10	61	2.33	8
f. Less employment positions	34	2.79	13	11	1.64	12	61	1.81	12
g. High participation costs	34	3.35	5	11	3.27	2	61	3.53	3
h. High project costs	34	3.03	11	11	2.18	9	61	2.43	7
i. A great deal of management time spent in contract transaction	34	3.29	6	11	2.55	5	61	3.86	1
j. Lack of experience and appropriate skills	33	3.27	8	11	3.45	1	61	2.78	5
k. Confusion over government objectives and evaluation criteria	34	3.41	4	11	3.00	3	61	2.81	4
l. Excessive restrictions on participation	34	3.29	7	11	2.55	6	61	2.32	9
m. Lengthy delays in negotiation	33	3.45	2	11	2.91	4	61	3.68	2

Source: Cheung and Chan 2009b; 2011a; Cheung et al. 2010c (with permission from Professor George Zillante, Editor of Proceedings of the 34th Australasian Universities Building Education Association Annual Conference, Adelaide, the American Society of Civil Engineers and Emerald Group Publishing Ltd).

N = Number of survey respondents

Another observation is that 'Less employment positions' and 'Reduce the project accountability' were both ranked within the bottom three of the rankings for all three jurisdictions. The respondents shared the same views on the negative factors they believed to be of less threat. The low ranking of 'Less employment positions' has shown that employment has not been affected, irrespective of how projects are procured. The main purpose of introducing PPP projects is not to 'Reduce the project accountability'; hence it was logical that all respondents perceived that this negative factor was less significant. Therefore these two negative factors were common for PPP projects irrespective of the geographical differences.

For the negative factors rated by respondents in Hong Kong the mean values ranged from 2.79 to 3.82. The variation in responses was 1.03. On the other hand, in Australia and the UK it was found that the mean values obtained ranged from 1.36 to 3.45 and 1.71 to 3.86 respectively. The variations in responses were 2.09 for Australia and 2.15 for the UK. Both variations were higher than that for Hong Kong. It was also found that in general these negative factors were rated higher by Hong Kong respondents. It can thus be interpreted that the Australian and British respondents found that these negative factors were less of a challenge. This finding is logical; as discussed previously Australia and the UK are much more experienced in delivering PPP projects compared to Hong Kong.

Similarly to the rating of the attractive factors, the respondents were also asked to rate the thirteen negative factors according to a Likert scale from 1 – 5 (1 = least important and 5 = most important), therefore a value above 3 would represent that the negative factor is of importance. The results show that in Hong Kong there were two negative factors below a score of 3. On the other hand in Australia and the UK there were ten and eleven respectively below 3. Again this consolidates the conclusion that the Australians and British are much more confident at conducting PPP projects, hence they are less conservative. The two negative factors ranked below 3 for Hong Kong were the ones discussed previously that were ranked low by all three sets of respondents. These negative factors were 'Less employment positions' and 'Reduce the project accountability', which both scored only 2.79.

In addition, on top of those factors the respondents were asked to rate, they were also given the opportunity to add others which would be of importance, but they did not do so.

Agreement of the survey respondents

As shown in Table 5.3, the Kendall's coefficient of concordance (W) for the rankings of attractive factors was 0.071 and 0.325 for Hong Kong and Australia respectively. The computed Ws were significant with $p = 0.008$ and 0.000 respectively. As the number of attributes considered was above seven, the chi-square value would be referred to rather than the W value. According to the degree of freedom, the critical value of chi-square was 23.680 for both groups (Hong Kong and Australia) the computed chi-square values were all above the critical value of chi-square (29.907 and 50.076 respectively). Therefore the assessment by the respondents within each group on their rankings of attractive factors is proved to be consistent. This finding ensures that the completed questionnaires were valid for further analysis.

Table 5.3 Results of Kendall's concordance analysis for the attractive factors of PPP

	Hong Kong	Australia
Number of survey respondents	30	11
Kendall's coefficient of concordance (W)	0.071	0.325
Chi-square value	29.907	50.076
Critical value of chi-square	23.680	23.680
Degree of freedom (df)	14	14
Asymptotic significance	0.008	0.000

Source: Cheung and Chan 2011a; Cheung *et al.* 2010c (with permission from the American Society of Civil Engineers and Emerald Group Publishing Ltd).

Note: Only 30 out of 34 responses from Hong Kong were suitable for subsequent statistical analyses.

Table 5.4 shows the Kendall's concordance analysis for the negative factors of PPP. The respective *W* for Hong Kong and Australia was 0.094 and 0.323. The number of attributes was also above seven, hence the chi-square value was referred to. The critical value of chi-square was 21.030 for both groups. The computed chi-square values were both higher at 35.968 and 42.591 for Hong Kong and Australia respectively. Hence the assessment by the respondents within each group on their rankings of negative factors is proved to be consistent. And this finding also ensures that the completed questionnaires were valid for further analysis.

The suitability of adopting PPP

With the identification of attractive and negative factors of PPP, these could be identified as checklists for assessing the suitability and/or feasibility of using PPP. If the attractive factors are prevailing in a given project scenario, the use of PPP

Table 5.4 Results of Kendall's concordance analysis for the negative factors of PPP

	Hong Kong	Australia
Number of survey respondents	32	11
Kendall's coefficient of concordance (W)	0.094	0.323
Chi-square value	35.968	42.591
Critical value of chi-square	21.030	21.030
Degree of freedom (df)	12	12
Asymptotic significance	0.000	0.000

Source: Cheung and Chan 2011a; Cheung *et al.* 2010c (with permission from and the American Society of Civil Engineers and Emerald Group Publishing Ltd).

Note: Only 32 out of 34 responses from Hong Kong were suitable for subsequent statistical analyses will be more positive. Conversely, if the negative factors are dominant PPP might be considered as unsuitable. Chapter 7 demonstrates further how this model can be applied for the Hong Kong Zhuhai Macau Bridge case study.

Chapter summary

This chapter presents the findings of an empirical questionnaire survey undertaken in Hong Kong, Australia and the UK to study the attractive and negative factors of conducting PPP projects. The survey respondents were asked to rate fifteen attractive factors and thirteen negative factors. The results gained from these three administrative regions were analysed and compared.

The results found that the top three attractive factors in Hong Kong were (1) Provide an integrated solution (for public infrastructure/services); (2) Facilitate creative and innovative approaches; and (3) Solve the problem of public sector budget restraint. These could be interpreted as situations where the use of PPP would be suitable. Similar results were found in the survey conducted in Australia. Efficiency-related factors appeared to be more attractive to the respondents from Hong Kong and Australia, whereas in the UK economic-related factors were rated higher. The finding for Hong Kong coincides with the fact that the local government has been enjoying a budget surplus in recent years, and has therefore not been pressured for delivering public projects with their own financial reserves.

The top three negative factors ranked by the respondents from Hong Kong were: (1) Lengthy delays because of political debate; (2) Lengthy delays in negotiation; and (3) Very few schemes have actually reached the contract stage (aborted before contract). These could be interpreted as situations where the use of PPP would be less desirable. The top negative factor was ranked of mediocre importance by the Australians and British, showing that they do not face the same concerns as Hong Kong. In Hong Kong this negative factor has been shown to be a problem, as demonstrated by the West Kowloon Cultural District project which was delayed mainly due to political debate. The second negative factor ranked in Hong Kong was also ranked high by the Australians and British. This negative factor was therefore seen to be important irrespective of the geographical differences and could be considered a negative factor specifically for PPP projects. The third negative factor was ranked bottom by the Australians and British, showing a high level of controversy to the ranking in Hong Kong. This factor appears to be more of a concern to the Hong Kong respondents. Some delayed projects, as discussed previously, are causes of the low confidence experienced. Australia and the UK, on the other hand, are much more experienced at conducting PPP projects, and hence more confident with this type of procurement.

6 Implementing public–private partnership projects

Introduction

This chapter studies the reasons, success factors and value for money measures behind public–private partnership (PPP) projects.

Reasons for implementing PPP projects

This section presents the findings of a study to investigate the reasons for implementing PPP projects (Cheung *et al.* 2009a). The same questionnaire survey described in Chapter 5 was used to obtain data. The survey respondents were asked to rate the importance of nine identified reasons for implementing PPP projects.

The reasons for implementing PPP projects were assessed from the different perspectives of the respondent groups in Hong Kong, Australia and the UK (results obtained by Li (2003) from his survey). The means for each administrative system were calculated and ranked in descending order of importance, as shown in Table 6.1.

As in Li's questionnaire, a total of nine reasons for implementing PPP projects were rated by the respondents. The top three reasons ranked in Hong Kong included:

1　Private incentive;
2　Economic development pressure demanding more facilities;
3　High quality of service required.

The top reason for implementing PPP projects ranked by respondents from Hong Kong was 'Private incentive'. Obviously practitioners around the world can foresee the advantages of involving the private sector in conducting public works projects. The private sector can add value to these projects in many ways such as financially, via expertise, innovation, risk sharing and above all motivation. This finding has indicated that the Hong Kong respondents felt that the main reason for implementing public works projects by PPP is to acquire the added value from the private sector. In Australia and the UK this reason for implementing PPP projects was ranked lower at fourth and ninth place respectively, indicating that those respondents did not feel so strongly about involving the private sector for its added value.

Table 6.1 Mean scores and rankings of the reasons for implementing PPP projects

Reasons	Hong Kong			Australia			United Kingdom (Li 2003)		
	N	Mean	Rank	N	Mean	Rank	N	Mean	Rank
a. Economic development pressure demanding more facilities	33	3.48	2	11	3.64	2	61	3.34	2
b. Political pressure	33	2.79	9	11	2.45	8	61	3.24	4
c. Social pressure of poor public facilities	33	2.88	8	11	3.09	5	61	3.12	5
d. Private incentive	32	3.56	1	11	3.09	4	61	2.57	9
e. Shortage of government funding	33	3.24	6	11	2.64	7	61	3.9	1
f. Inefficiency because of public monopoly and lack of competition	33	3.33	4	11	3.09	3	61	2.98	6
g. High quality of service required	33	3.42	3	11	3.91	1	61	2.7	7
h. Avoid public investment restriction	33	2.97	7	11	2.18	9	61	3.31	3
i. Lack of business and profit-generating skill in the public sector	32	3.31	5	11	2.82	6	61	2.62	8

Source: Cheung *et al.* 2009a (with permission from Emerald Group Publishing Ltd).

N = Number of survey respondents

Ranked second by respondents in Hong Kong, Australia and the UK was 'Economic development pressure demanding more facilities'. The similar ranking pattern across the three survey groups suggests that the importance of this reason for implementing PPP projects is applicable irrespective of geographical differences. Hence all survey respondents felt that PPP projects are implemented due to economic pressure to provide more public facilities. The similar ranking pattern could also be a reflection of the real life situation that the survey respondents have observed. In Hong Kong particularly, there has been a growing phase of rapid infrastructure development, for which the government has opted to use PPP. These projects include the Shatin to Central rail link and the Kwun Tong rail extension. The new metro line will consist of nine stations. Construction will start in 2010 and the two phases of the line will be completed by 2015 and 2019 (Information Services Department 2008a).

The third reason for implementing PPP projects ranked by respondents from Hong Kong was 'High quality of service required'. Being an international city, maintaining high quality in services is important. In Australia and the UK this reason for implementing PPP projects was ranked first and seventh respectively. The findings show that the Australians felt similarly but the British ranked this reason for implementing PPP projects much lower. However, the survey with the British respondents was conducted a few years ago, and it is anticipated that with the increasing number of projects undertaken for the Olympics in 2012, the respondents might have a different view if this survey was conducted today.

In Australia, the respondents ranked 'Inefficiency because of public monopoly and lack of competition' third. Due to the size, complexity, challenges and long concession period of PPP projects, they tend to be limited to being conducted only by very large private sector companies. These companies will normally possess sufficient finance, expertise and skills to implement PPP projects. Therefore those who are not involved with the PPP process may feel that public monopoly and lack of competition exists. This occurrence is often partially true, but then only those capable parties will possess the power to participate with PPP projects.

Ranked first by British respondents was 'Shortage of government funding'. One of the main reasons for the rise of PPP/Private Finance Initiative (PFI) projects in the UK was financial resources from the private sector. The PPP/PFI method was first adopted at a time when the British government was struggling to provide for public facilities and services (Zhang 2001). By involving the private sector the government was able to continue delivering public infrastructure. As a result a heavy emphasis on finance has always been associated with PPP/PFI projects especially in the early days of implementation. Along with other benefits as a result of involving the private sector, finance is, however, often not the only element considered when delivering public projects these days.

Third in the UK rank was 'Avoid public investment restriction'. Similar to the reason discussed previously, this reason has a strong emphasis on the financial element of the project. Again it must be considered that the survey conducted with British respondents was carried out a few years ago. It is likely that when the British government was still in a tight budgetary condition it would also be more likely to enforce more budgetary restrictions before approving projects. Hence it is unsurprising for this reason to be ranked highly by the British respondents.

The mean values of the reasons for implementing PPP projects as rated by Hong Kong respondents ranged from 2.79 to 3.56. This observation has reflected that the variation in their responses is relatively small, only 0.77 for Hong Kong. In Australia and the UK the means ranged from 2.18 to 3.91 and 2.57 to 3.90, respectively. The corresponding differences in means were 1.73 and 1.33 respectively. The differences in means were much higher for the survey conducted in Australia and the UK compared to Hong Kong.

As the respondents were asked to rate the nine reasons for implementing PPP projects according to a Likert scale from 1 to 5 (1 = least important and 5 = most important), a value above 3 would represent that the reason for implementing PPP projects is of importance. Amongst the reasons for implementing PPP projects only three were ranked below 3 in the Hong Kong rankings. These reasons for implementing PPP projects were 'Political pressure', 'Social pressure of poor public facilities', and 'Avoid public investment restriction' which scored 2.79, 2.88 and 2.97 respectively.

For Australia and the UK, each had four reasons for implementing PPP projects rated below 3. In Australia, two of these were the same as those for Hong Kong ('Political pressure' and 'Avoid public investment restriction' with scores of 2.45 and 2.18 respectively). The other two in Australia were 'Shortage of government funding' and 'Lack of business and profit-generating skill in the public sector' which scored 2.64 and 2.82 respectively. On the other hand, in the UK none were the same as those in Hong Kong but one was the same as Australia ('Lack of business and profit-generating skill in the public sector' which scored 2.62). The other three reasons in the UK were 'Private incentive', 'Inefficiency because of public monopoly and lack of competition' and 'High quality of service required' which scored 2.57, 2.98 and 2.7, respectively.

The reason for implementing PPP projects 'Political pressure' was rated low by respondents in both Hong Kong and Australia. Hood and McGravey (2002) claimed that PPP development would remain a major political issue. Relatively speaking, Hong Kong and Australia have less history of PPP implementation compared to the UK. Also, they faced less political pressure when the concept was first introduced, as the practice has been well documented in other developed countries (such as the UK) and the political influence of trade unions is minimal. Hence this reason for implementing PPP projects was not rated highly.

Also rated lowly by respondents from Hong Kong and Australia was 'Avoid public investment restriction'. Again this reason for implementing PPP projects was not rated highly as both groups of survey respondents did not believe that the public were under heavy investment restrictions.

Rated low by only the Hong Kong respondents was 'Social pressure of poor public facilities'. The Hong Kong respondents did not feel that the government has been under pressure from society. Hence they rated this reason for implementing PPP projects lowly. This finding could imply that the respondents felt happy about the current standard of public facilities in Hong Kong.

Rated lowly by the Australian respondents only was 'Shortage of government funding'. Although financial drive may once have been the main reason for involving private sector participation, this is no longer the case. In Australia, the state governments have noticed the benefits associated with implementing PPP projects and have developed a more revolutionary process. The state governments are capable of delivering these services themselves but instead choose to involve the private sector to achieve added value for particular public projects.

Rated lowly by the Australian and British respondents was 'Lack of business and profit-generating skill in the public sector'; again, the Australians and the British have a much longer history in implementing PPP projects hence their skills in this area are much more advanced. As a result the public sector has acquired sufficient experience and competency to deliver these projects well. Therefore the respondents felt that incapability of the public sector to deliver public projects was not the case.

The British respondents rated 'Private incentive' lowly. This contradicts the finding achieved from the Hong Kong respondents. The public sector of the UK is already well experienced at conducting PPP projects, but they realise their job is to deal with the administrative procedures rather than act as the developer.

Also rated lowly in the UK survey was 'Inefficiency because of public monopoly and lack of competition' and 'High quality of service required'. Again, the experience of the public sector implies that these can be achieved without involving the private sector.

In addition, on top of those factors the respondents were asked to rate, they were also given the opportunity to add others which would be of importance, but they did not do so.

The findings have shown that in general those reasons ranked highly by respondents from Hong Kong and Australia focused on improving the overall performance of public projects, whereas those that were rated highly by the British respondents focused on the financial aspect of the projects. Ranked in the top three by Hong Kong respondents were 'Private incentive', 'Economic development pressure demanding more facilities' and 'High quality of service required'. In Australia and the UK both groups of respondents also ranked their second reason the same as Hong Kong. In addition, the Australians also ranked the third reason in Hong Kong first, and 'Inefficiency because of public monopoly and lack of competition' third. In the UK the first and third reasons ranked by the respondents was 'Shortage of government funding', and 'Avoid public investment restriction', respectively.

The reason 'Private incentive' was attractive because of the added value which could be applied to public works projects by the private sector. One of the main reasons to adopt PPP is that a public works project can benefit from the private sector's expertise, innovation, motivation and experience. Similarly for many governments around the world, 'Economic development pressure demanding more facilities' is common. Even though governments such as Hong Kong's are able to finance their own projects, there are also other areas in society which they need to support. So by using money from the private sector, governments can utilise their resources much more effectively. In international cities particularly, 'High quality of service required' to maintain their status and competition is common. The size and complexity of PPP projects often limit only certain large private sector parties therefore 'Inefficiency because of public monopoly and lack of competition' is often seen. Many governments first started to implement PPP projects due to 'Shortage of government funding'. Similarly, when the government is under tight budget controls implementing PPP projects could also 'Avoid public investment restriction'.

Factors contributing to successful PPP projects

This section presents the findings of a study to investigate the factors contributing to successful PPP projects (Chan *et al.* 2010; Cheung *et al.* 2012). The same questionnaire survey was used to obtain data. The survey respondents were asked to rate eighteen factors which contribute to delivering successful PPP projects.

The factors that contribute to the success of PPP projects were assessed from different perspectives of the respondent groups in Hong Kong, Australia and the UK (results obtained by Li (2003) from his survey). The means for each administrative system were calculated and ranked in descending order of importance as shown in Table 6.2.

Table 6.2 Mean scores and rankings for the factors that contribute to the success of PPP projects

Success factors	Hong Kong			Australia			United Kingdom (Li 2003)		
	N	Mean	Rank	N	Mean	Rank	N	Mean	Rank
a. Stable macro-economic condition	34	3.85	4	11	4.18	12	61	3.19	15
b. Favourable legal framework	34	4.06	1	11	4.27	7	61	3.63	9
c. Sound economic policy	34	3.74	7	11	4.09	13	61	3.19	13
d. Available financial market	34	3.71	8	11	4.18	11	61	4.04	3
e. Multi-benefit objectives	34	3.50	16	10	4.20	10	61	3.19	14
f. Appropriate risk allocation and risk sharing	34	3.85	5	11	4.64	2	61	4.05	2
g. Commitment and responsibility of public and private sectors	34	3.97	2	11	4.91	1	61	3.98	4
h. Strong and good private consortium	34	3.91	3	11	4.64	3	61	4.11	1
i. Good governance	34	3.68	10	11	4.45	4	61	3.72	8
j. Project technical feasibility	34	3.56	15	11	4.36	5	61	3.79	6
k. Shared authority between public and private sectors	34	3.41	18	10	3.70	16	61	2.98	17
l. Political support	34	3.76	6	11	4.27	6	61	3.56	11
m. Social support	34	3.44	17	11	3.36	17	61	2.81	18
n. Well organised and committed public agency	34	3.65	12	11	4.27	8	61	3.74	7
o. Competitive procurement process (enough potential bidders in the process)	34	3.68	9	11	4.27	9	61	3.37	12
p. Transparent procurement process (process is made open and public)	33	3.67	11	11	4.09	14	61	3.6	10
q. Government involvement by providing guarantee	34	3.62	14	10	2.40	18	61	3.16	16
r. Thorough and realistic assessment of the cost and benefits	34	3.65	13	11	4.00	15	61	3.95	5

Source: Cheung et al. 2012 (with permission from Emerald Group Publishing Ltd).

* N = Number of survey respondents

Eighteen success factors for adopting PPP were rated by the respondents. Figure 6.1 illustrates the relationship of the top five success factors ranked in Hong Kong with their ranking positions in Australia and the UK. These success factors included:

1   Favourable legal framework;
2   Commitment and responsibility of public and private sectors;
3   Strong and good private consortium;
4   Stable macro-economic condition;
5   Appropriate risk allocation and risk sharing.

The success factor ranked top by respondents from Hong Kong was 'Favourable legal framework'. On the other hand, respondents from Australia and the UK ranked this success factor of medium importance only, at seventh and ninth position respectively. This finding implies that the Australian and British respondents were not particularly concerned about their existing legal framework, which is already well established to handle PPP projects.

On the contrary, respondents in Hong Kong felt that the legal framework is the most important success factor. As mentioned by the National Treasury PPP Unit of South Africa (2007), an independent, fair and efficient legal framework is a key factor for successful PPP project implementation. Sufficient legal resources at reasonable costs should be available to deal with the amount of legal structuring and documentation required. A transparent and stable legal framework would help to make the contracts and agreements bankable. An adequate dispute resolution system would help to ensure stability in the PPP arrangements. Appropriate governing rules, regulations and reference manuals related to PPP have been well established in some developed countries (for example, the UK, Australia, Canada and South Africa) to facilitate the effective application of PPP procurement.

The success factor ranked second by Hong Kong respondents was 'Commitment and responsibility of public and private sectors'. This success factor was also ranked importantly by the Australians and British at first and fourth place respectively. This success factor was ranked high by all respondent groups irrespective of geographical location. To secure the success of PPP projects, both the public and private sectors should bring their complementary skills and commit their best resources to achieve a good relationship (National Audit Office 2001).

Ranked third by respondents in Hong Kong and Australia was 'Strong and good private consortium'. Respondents from the UK felt even more strongly about the importance of this success factor and ranked it top. This finding again has shown that this success factor is seen to be highly important to the success of PPP projects irrespective of geographical locations. The government in contracting out the PPP projects should ensure that the parties in the private sector consortium are sufficiently competent and financially capable of taking up the projects. This suggests that private companies should explore other participants' strengths and weaknesses and, where appropriate, join together to form consortia capable of synergising and exploiting their individual strengths. Good relationships among partners are also critical because they all bear relevant risks and gain benefits from the co-operation

(Abdul-Rashid *et al.* 2006; Birnie 1999; Corbett and Smith 2006; Jefferies *et al.* 2002; Kanter 1999; Tam *et al.* 1994; Tiong 1996; Zhang 2005a).

The success factor ranked fourth by respondents from Hong Kong was 'Stable macro-economic condition'. Respondents from Australia and the UK ranked this success factor relatively low amongst the eighteen success factors rated. In these countries this success factor was ranked only twelfth and fifteenth respectively. This success factor was therefore seen as quite important in Hong Kong but rather unimportant in Australia and the UK.

In a stable macro-economic environment the market is more predictable, hence lowering risks such as the interest rate, exchange rate, employment rate, inflation rate, and so on. It is very important to reduce risks and enable a reasonable investment return for private investors, especially in the emerging PPP market of Hong Kong. For projects where the major source of revenue to the private sector is generated from direct tariffs levied on users, there are revenue risks that are beyond the control of the private sector, such as future usage level and permitted tariff charges.

There may also be unforeseen risks arising during the course of the project life. To ensure project economic viability, the government may consider some forms of government guarantees, joint investment funding, or supplemental periodic service payments to allow the private sector to cover the project costs and earn reasonable profits and investment returns. At the same time, the government should take due consideration of the private sector's profitability requirements in order to have stable arrangements in PPP projects. Alternative sources of income and financing, like property development opportunities along the railway, can be sought to bridge the funding gap for private investors (Abdul-Rashid *et al.* 2006; Corbett and Smith 2006; Li *et al.* 2005c; Nijkamp *et al.* 2002; Qiao *et al.* 2001; Tam *et al.* 1994; Tiong 1996; Zhang 2005b).

The success factor ranked fifth by Hong Kong respondents was 'Appropriate risk allocation and risk sharing'. Respondents from Australia and the UK both ranked this success factor second, showing again that irrespective of geographical differences all groups of respondents ranked this success factor as important. However, the findings showed that the respondents from Hong Kong ranked this success factor slightly below the others. This could be because Hong Kong has had experience with different procurement systems that entail different risk allocation models, thereby making this success factor relatively less critical in terms of its contribution to project success.

A core principle in PPP arrangement is the allocation of risk to the party best able to manage and control it (Efficiency Unit 2003). Logically, a government would prefer to transfer risks associated with asset procurement and service delivery to the private sector participants, who are generally more efficient and experienced in managing them. But a government should be able to take up risks that are beyond the control of private sector participants. In all cases, a government should ensure there are measures in place to manage risk exposure rather than leaving it open to the private sector. Likewise before committing to the projects, the private sector participants should fully understand the risks involved and should be prudent in pricing and managing the risks appropriately (Grant 1996; Qiao *et al.* 2001; Zhang 2005c).

The results also found that the success factors 'Shared authority between public and private sectors' and 'Social support' were ranked in the bottom three by all groups of respondents. These success factors were therefore seen to be least important compared to the others. Although no explanation can be provided for why these success factors were ranked lower, it can be assumed that the other success factors were seen to be more important. This perception was shown to be true for all survey locations.

The mean values for the success factors as rated by Hong Kong respondents ranged from 3.41 to 4.06. This observation has reflected that the variation in their responses are relatively small, only 0.65 for Hong Kong. In Australia and the UK the means ranged from 2.40 to 4.91 and 2.81 to 4.11 respectively. The corresponding differences in means were 2.51 and 1.30 respectively. The differences in means were shown to be much higher for the survey conducted in Australia and the UK compared to Hong Kong. This finding shows that the Hong Kong respondents rated the eighteen success factors much more similarly, whereas in Australia and the UK the respondents showed a much larger variation.

As the respondents were asked to rate the eighteen success factors according to a Likert scale from 1 to 5 (1 = least important and 5 = most important), a value above 3 would represent that the success factor is of importance. Amongst the success factors none were ranked below 3 in the Hong Kong ranking. In Australia and the UK only one and two success factors respectively were ranked below a score of 3. Therefore, all three groups of respondents concurred that the identified factors were important.

In addition, on top of those factors the respondents were asked to rate, they were also given the opportunity to add others which would be of importance, but they did not do so.

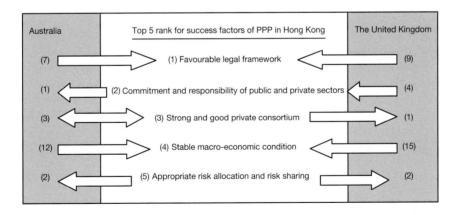

Figure 6.1 Rank relationship between Hong Kong, Australia and the UK for success factors of PPP (Cheung *et al.* 2012) (with permission from Emerald Group Publishing Ltd)

This section has analysed the perceptions of respondents from Hong Kong, Australia and the UK on the importance of eighteen factors contributing to successful delivery of PPP projects. The ranking in Hong Kong showed that the top five success factors were: (1) Favourable legal framework; (2) Commitment and responsibility of public and private sectors; (3) Strong and good private consortium; (4) Stable macro-economic condition; and (5) Appropriate risk allocation and risk sharing.

The top success factor according to Hong Kong respondents was ranked with medium importance by respondents from the other two groups, implying that their legal frameworks are already well developed to cater for PPP projects, making them less concerned about the existing system.

The second, third and fifth success factors were ranked highly by all three groups of respondents, indicating that these success factors were applicable to delivering successful PPP projects irrespective of their geographical locations.

The success factor ranked fourth by Hong Kong respondents was ranked lowly by the other respondent groups. In Australia and the UK there is a well-established stable macro-economic environment, so the market is much more predictable. Hong Kong, on the other hand, has experienced dramatic changes since the British-to-Chinese handover in 1997, hence they are still adjusting to the changes and the market is therefore not as stable. As a result a stable macro-economic condition was rated much higher by the Hong Kong respondents. In general, all three groups of respondents concurred that the identified factors were important.

Enhancing value for money in PPP projects

This section presents the findings of a study investigating the measures that enhance value for money (VFM) in PPP projects (Cheung *et al.* 2009b). The same questionnaire survey was used to obtain data. The survey respondents were asked to rate the importance of eighteen VFM measures in PPP projects.

The VFM measures in PPP were assessed from different perspectives of the respondent groups from Hong Kong, Australia and the UK (results obtained by Li (2003) from his survey). The means for each administrative system were calculated and ranked in descending order of importance as shown in Table 6.3.

Eighteen VFM measures in PPP were rated by the respondents. Figure 6.2 illustrates the relationship of the top five VFM measures ranked in Hong Kong with their ranking positions in Australia and the UK. These VFM measures comprised:

1   Efficient risk allocation (allocating the risk to the party best able to manage it);
2   Output-based specification;
3   Competitive tender;
4   Private management skill;
5   Private sector technical innovation.

The VFM measure ranked top by the Hong Kong respondents was 'Efficient risk allocation (allocating the risk to the party best able to manage it)'. This VFM measure was also ranked top by the Australians and highly at second place by the

Table 6.3 Mean scores and rankings for the VFM measures in PPP projects

VFM measures	Hong Kong			Australia			United Kingdom (Li 2003)		
	N	Mean	Rank	N	Mean	Rank	N	Mean	Rank
a. Competitive tender	34	3.91	3	11	4.27	6	61	3.5	6
b. Efficient risk allocation (allocating the risk to the party best able to manage it)	33	4.18	1	11	4.55	2	61	4.02	1
c. Risk transfer (transferring a substantial amount of risk from the public to the private)	34	3.59	8	11	2.73	17	61	3.57	5
d. Output-based specification	34	3.91	2	11	4.27	5	61	3.91	2
e. Long-term nature of contracts	34	3.65	7	11	4.18	7	61	3.78	3
f. Improved and additional facilities to the public sector	34	3.35	12	11	4.00	11	61	3.16	13
g. Private management skill	34	3.82	4	11	4.27	4	61	3.41	7
h. Private sector technical innovation	33	3.82	5	10	4.50	3	61	3.28	9
i. Optimal use of asset/facility and project efficiency	34	3.68	6	10	4.70	1	61	3.31	8
j. Early project service delivery	34	3.35	11	11	4.00	10	61	3.72	4
k. Low project life cycle cost	34	3.47	10	11	4.00	9	61	3.24	11
l. Low shadow tariffs/tolls	34	2.82	18	10	3.30	13	61	2.49	17
m. Level of tangible and intangible benefits to the users	34	3.00	16	11	4.00	8	61	2.83	15
n. Environmental consideration	34	2.97	17	11	2.73	16	61	2.38	18
o. Profitability to the private sector	34	3.18	13	10	3.00	15	61	2.84	14
p. 'Off the public sector balance sheet' treatment	34	3.15	14	11	2.36	18	61	3.23	12
q. Reduction in disputes, claims and litigation	34	3.09	15	11	3.18	14	61	2.81	16
r. Nature of financial innovation	34	3.56	9	11	3.73	12	61	3.25	10

Source: Chan et al. 2008f; Cheung et al. 2009b (with permission from Emerald Group Publishing Ltd and International Council for Research and Innovation in Building and Construction).

N = Number of survey respondents

British, showing that its importance in PPP projects is applicable irrespective of geographical locations.

It is essential for the public client and the private bidders to evaluate all of the potential risks throughout the whole project life. Public and private sector bodies must place particular attention on the procurement process while negotiating contracts for PPP to ensure a fair risk allocation between them. Systematic risk management allows early detection of risks and encourages the PPP stakeholders to identify, analyse, quantify and respond to the risks, as well as take measures to introduce risk mitigation policies (Akbiyikli and Eaton 2004). A fundamental principle is that risks associated with the implementation and delivery of services should be allocated to the party best able to manage the risk in a cost effective manner.

Second in the Hong Kong and Australia rank was 'Output-based specification'. This VFM measure was also ranked highly by the British at fifth. Besides the top VFM measure ranked by Hong Kong discussed previously, this was the only one also ranked highly by all three administrative regions, indicating again that this VFM measure is applicable to PPP projects irrespective of geographical differences.

Clear specifications can be used to quantify the resources required for a project. When project specifications are more difficult to define, the costs that it may incur are also hard to quantify and control. Therefore clearly defined, output-based specifications can help the government to monitor the private sector's performance. The private party can also feel more confident in achieving targets and keeping control of the project flow in order to enhance their profit margins. Output-based specifications can also help the government to use the public sector comparator more effectively in quantifying whether VFM is reached by procuring projects via PPP. Some may feel that output-based specifications define too much of the project to allow for private sector innovation, but, for example, a two-lane tunnel can still leave plenty of room for added value from the private sector.

The Hong Kong respondents ranked 'Competitive tender' third. In Australia and the UK this VFM measure was ranked with medium importance only by both, at sixth position. This VFM measure was ranked highly in Hong Kong reflecting the respondents' views of the actual situation of procuring projects.

Hong Kong has only a limited number of contractors who are able to handle large public works projects. Therefore it is often the same groups of contractors who are successful at winning these bids. Those slightly smaller local companies are often unable to compete with the larger local companies. International companies based in Hong Kong may not always wish to spend their resources in Hong Kong. Hence a revolving situation has developed, where there are often few bids received from the private sector. As a result these projects tend to be awarded to the same groups of people. Therefore a situation evolves where the fewer competitors in the tendering process the more difficult it is to achieve VFM in PPP projects.

In a more competitive bidding environment, the private sector will try all measures to improve their designs in every aspect. This is particularly so with VFM, as one of the main reasons that the public sector opt for PPP is to achieve VFM in public works projects. This would therefore be a key reason for a government to choose a particular private party. In a bidding environment that has few competitors the private sector does not need to try so hard to win the contracts, hence VFM may not always be achieved.

Ranked fourth in Hong Kong and Australia was 'Private management skill'. This VFM measure was ranked slightly lower by the British at seventh position. The British are more experienced in conducting PPP projects hence many of the private sector companies are already equipped with the necessary skills to handle them. On the other hand, in Hong Kong particularly, many private companies are not experienced with handling PPP projects and are therefore not equipped with the necessary management skills. The capability of the private sector can determine the successfulness of the PPP project. The success of a PPP project is often associated with its degree of VFM that can be achieved.

The fifth VFM measure in Hong Kong was 'Private sector technical innovation'. This was ranked slightly higher by the Australians but lower by the British at third and ninth position respectively. This VFM measure is similar to 'Private management skill', in that it relies on the capability of the private party. Obviously the ability of the private party will determine how successful the PPP project can become in terms of VFM. Then again, VFM is the main incentive for governments around the world to involve the private sector in procuring public works projects.

The mean values for the VFM measures as rated by Hong Kong respondents ranged from 2.82 to 4.18. This observation shows that the variation in their responses is relatively small, only 1.36 for Hong Kong. In Australia and the UK the means ranged from 2.36 to 4.70 and 2.38 to 4.02 respectively. The corresponding differences in means were 2.34 and 1.64 respectively. The differences in means were slightly higher for the survey conducted in Australia and the UK compared to Hong Kong. This finding shows that the Hong Kong respondents rated the eighteen VFM measures much more similarly, whereas in Australia and the UK the respondents showed a slightly larger variation.

As the respondents were asked to rate the eighteen VFM measures according to a Likert scale from 1 to 5 (1 = least important and 5 = most important), a value above 3 would represent that the VFM measure is of importance. Amongst the VFM measures only two were ranked below 3 in the Hong Kong ranking. These VFM measures were 'Low shadow tariffs/tolls' and 'Environmental consideration', which scored 2.82 and 2.97 respectively. In Australia and the UK, three and five VFM measures were rated below 3 respectively. Similarly to Hong Kong, the other respondent groups also rated 'Environmental consideration' below a score of 3 with scores of 2.73 and 2.38 respectively. This finding showed that environment-related issues showed the least effect towards enhancing VFM according to all groups of survey respondents.

In addition, on top of those factors the respondents were asked to rate, they were also given the opportunity to add others which would be of importance, but they did not do so.

This section has discussed the VFM measures rated by survey respondents from Hong Kong, Australia and the UK. The results showed that there were two VFM measures that were ranked highly by all groups of survey respondents.

The first of these which was ranked top in Hong Kong was 'Efficient risk allocation (allocating the risk to the party best able to manage it)'. Appropriate risk

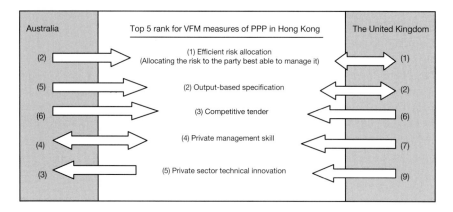

Figure 6.2 Rank relationship between Hong Kong, Australia and the UK for VFM measures of PPP (Cheung *et al.* 2009b) (with permission from Emerald Group Publishing Ltd)

allocation, so that risks are assigned to the party best able to manage it, is believed to reduce the problems encountered in a project. As a result VFM is enhanced because fewer risks occur in the project life.

The second VFM measure ranked highly by all was 'Output-based specification'. A clearly defined output-based specification enables the milestones and activities in a project to be much more predictable compared to one without, hence the effect towards VFM is larger.

Ranked third in Hong Kong was 'Competitive tender'. This measure can create VFM when competitive tendering exists. The more competition in the tendering process, the more the private sector will try to offer a better package overall for the public sector. In Hong Kong, unfortunately, there is limited competition between those companies that can handle PPP projects; hence the respondents felt that this VFM measure is relatively more important.

Ranked fourth and fifth in Hong Kong were 'Private management skill' and 'Private sector technical innovation'. Both of these VFM measures relate to the ability of the private sector. Obviously the better the private sector's ability, the more chance there is for them to enhance VFM. In Hong Kong the skill of the private sector in conducting PPP projects may not be as well-developed by experience as in Australia and the UK, hence the respondents felt strongly about these measures.

Chapter summary

This chapter has presented the findings from a questionnaire survey looking at the reasons for implementing PPP projects, the success factors of PPP projects and the VFM measures of PPP projects.

The findings showed that the top reason ranked by the survey respondents in Hong Kong was 'Private incentive'. Ranked second by all three groups of survey respondents was 'Economic development pressure demanding more facilities'. Third in Hong Kong and first in Australia was 'High quality of service required'. The reason 'Inefficiency because of public monopoly and lack of competition' was ranked third by the Australian respondents. And finally, ranked first and third by the British respondents was 'Shortage of government funding' and 'Avoid public investment restriction'. The rankings showed that in general those reasons rated highly in the UK focused on financial elements whereas those rated highly in Hong Kong and Australia were more related to the overall performance of improving public projects.

Amongst the top five success factors ranked by Hong Kong respondents, three were also ranked highly by the Australians and British. These success factors included: 'Commitment and responsibility of public and private sectors', 'Strong and good private consortium' and 'Appropriate risk allocation and risk sharing'. These success factors were therefore found to be important for contributing to successful PPP projects irrespective of geographical locations. Ranked top in Hong Kong but only with medium importance in the other surveyed jurisdictions was 'Favourable legal framework'. Also, ranked within the top five by Hong Kong respondents was 'Stable macro-economic condition', but this success factor was ranked lowly by the Australians and British.

The top five VFM measures ranked by the respondents from Hong Kong included: 'Efficient risk allocation (allocating the risk to the party best able to manage it)', 'Output-based specification', 'Competitive tender', 'Private management skill', and 'Private sector technical innovation'. The first and second of these VFM measures were also found to be ranked highly by the respondents from Australia and the UK, indicating that these were true irrespective of geographical differences. When the risks are handled well, fewer pitfalls are experienced and as a result VFM is more achievable. Hence an efficient risk allocation is vital in determining whether VFM can be achieved in PPP projects. A clear output-based specification can enable a more obvious project design and concept hence minimising the possibility of delivering the wrong product for the user. Therefore this measure is also important in determining whether VFM has been achieved for a PPP venture.

7 An evaluation model for assessing the suitability of public–private partnership

Introduction

This chapter presents an evaluation model which can be used for assessing the suitability of public–private partnership (PPP) projects by studying their attractive and negative factors (Cheung and Chan 2009a; 2011a). A questionnaire survey was conducted with industrial practitioners in Hong Kong (as described in Chapter 5). The respondents were requested to rank the importance of fifteen attractive factors and thirteen negative factors for adopting PPP. From the rankings, the relative weightings of each factor were derived. The weightings of these factors formed the basis for the evaluation model presented in this chapter. The Hong Kong–Zhuhai–Macau Bridge (HKZMB) was used to demonstrate how this evaluation model could be applied. From the list of attractive and negative factors the authors selected those which were foreseeable in the HKZMB project. By calculating the total weightings of each group of factors the result indicated that PPP would not be a suitable method for delivering this project. The evaluation model presented in this chapter can help both the public and private sectors to assess whether potential public projects are suited for PPP. Academics are also shown how their research work could be delivered to a wider audience and applied in more practical situations within the industry.

Development of the evaluation model

Step 1 Establish the weighting of attractive and negative factors

The weightings of attractive and negative factors were established via a questionnaire survey. Details of the questionnaire design and administration, selection criteria for questionnaire respondents, and background of questionnaire respondents have already been described in Chapter 5 of this book.

Step 2 Analyse the potential PPP project

A thorough analysis of the potential PPP project being considered should be conducted. Aspects of the project which should be studied include: history, development, future, parties involved, view of general public, preference of public and private sector, normal practice, advantages and disadvantages, political situation, timeframe, opportunities, obstacles, culture, and so on. These types of information can be sourced from newspapers, magazines, governmental reports and websites, studies conducted by researchers, private sector publications, interviews with parties involved or parties that would be affected, discussions with experts, questionnaire survey with the general public, and so on. The user of the evaluation model will match the project information available to the list of attractive and negative factors. For each factor the user will then assign a score for the likelihood that it would occur in the project being considered. The score will be given according to the same Likert scale used by the questionnaire respondents.

Step 3 Evaluate the decision for adopting PPP

The total score for the attractive and negative factor groups can be derived by the sum of multiplying the relative weighting of the factor (which is the mean score given by the questionnaire respondents) by the score of the factor (this is the score given by the user of the evaluation model). The total score can be expressed by the following formula:

TS = ΣWt × S
TS = Total score of factor group (attractive or negative factor group)
Wt = Weighting of individual factor within the factor group
S = Score assigned by user for individual factor within factor group

The total score of the group of attractive factors will be compared with that of the group of negative factors. The group of factors that scores the highest indicates the suitability of adopting PPP for the project being considered. For example, if the total score of the attractive factors is higher, PPP is the preferred option, whereas if the total score of the negative factors is more dominant then PPP is not recommended.

Weighting of the attractive and negative factors

The attractive and negative factors for adopting PPP were assessed by respondents from Hong Kong. The means for each factor were calculated and ranked in descending order of importance.

The suitability of using PPP for a high-profile case: the Hong Kong–Zhuhai–Macau Bridge

Background of the Hong Kong–Zhuhai–Macau Bridge

The proposed design

It is believed the HKZMB will enhance the economic development of Hong Kong, Macau and the Western Pearl River Delta region (Hong Kong Special

Administrative Region Government 2008). The new bridge is expected to significantly reduce the cost and time for both people and goods transportation between the regions. At the same time it is hoped that the project will increase the region's competitiveness. The construction of the bridge is expected to commence no later than 2010 (Hong Kong Special Administrative Region Government 2008). The estimated completion date is set for 2015 to 2016 (Chen and Lee 2008).

The initial estimated time of travel across the bridge is believed to be within 15 to 20 minutes and the total cost of the bridge will be approximately RMB37.4 billion (approximately US$6 billion on 11 December 2012, Yahoo! Finance 2012) (Mak 2008). The main bridge will be a 29.6km three-lane dual carriageway in the form of bridge tunnel structure, comprising an immersed tunnel of about 6.7 kilometres. Vehicle speeds are anticipated to be 100 kilometres per hour. A traffic flow of approximately 12,000–16,000 vehicles are expected per day (Hung 2008). The bridge will land on an artificial island off Gongbei on the west side, and another artificial island on the east which would be west of the Hong Kong boundary. According to the current proposed construction option, the connecting roads are about 12.6 kilometres on the Hong Kong side and 13.9 kilometres on the mainland side. The bridge will run across the Lingding Channel, the Tonggu Channel, the Qingzhou Channel, the Jiuzhou Port Channel and the Jianghai Channel (Transport and Housing Bureau 2008a).

The original PPP decision

The HKZMB project was originally suggested by the private sector, which is why PPP was the assumed delivery method from the very beginning. A PPP plan for the bridge was originally drawn up in early 2008. This plan was officially initiated by the three governments from Guangdong, Hong Kong and Macau in 2002. Under the PPP scheme, the three governments would be responsible only for construction of ports and connective parts of the bridge within the three sides, and its main part will be constructed by bids (Qiu 2008). Under the PPP arrangement, the bridge was to have a fifty-year concession period (Legislative Council 2008).

Another reason for the HKZMB to be delivered by the PPP model was the high project costs. By involving the private sector the governments would not need to take up the financial risks involved (Apple Daily 2008).

Changing from private financing to public financing

The idea for the HKZMB was first proposed by Mr Gordon Wu of the Hong Kong listed Hopewell Holdings Ltd in the 1980s (Kwok 2009). Mr Wu's original initiation of the project led minds to think that Hopewell Holdings would definitely be interested in participating in this mega infrastructure project, but it has been reported that more than twenty years after the idea was first proposed, the company no longer saw a business opportunity in the plan (Lam and Chan 2008). Other private sector companies felt the same; the private sector was no longer interested in this project as the business potential for them was not attractive.

Therefore, the decision for the financing model of the bridge was changed dramatically. In the 'Eighth AWCG Meeting' held in February 2008, it was still assumed that the project would be procured by PPP. The three governments agreed to take up the responsibility for construction and operation of the boundary-crossing facilities and the link roads to the bridge within their own territory. It was discussed that private investment would be invited for the main body of the bridge with the funding gap shared by the three governments according to construction needed in their own territories. In this arrangement Hong Kong would have covered 50 per cent of the difference, Guangdong 35 per cent and Macau 15 per cent. The decision showed that the governments were in favour of the PPP arrangement at the time (Transport and Housing Bureau 2008b).

But in an interview conducted with the Secretary for Transport and Housing Bureau also in February 2008, she was asked by reporters whether the PPP method would be adopted for the HKZMB. The Secretary responded that the project would be considered as a whole amongst the governments. Her response did not directly answer whether the project would be financed by the private sector or not (Transport and Housing Bureau 2008c).

Under some discussion and reviews of the studies that have been carried out such as on the traffic flow and bid price, it was realised that the governments would not be able to come up with an attractive economic package for the private sector to be interested (Ming Pao Newspaper 2008a). Finally a decision was made at the '11th Plenary of Hong Kong-Guangdong Co-operation Joint Conference' in August 2008. It was announced that the HKZMB would be funded jointly by the governments (Hong Kong Special Administrative Region Government 2008). It was confirmed that the bridge would be conducted using public money rather than private sector resources.

The preliminary proposed contribution from each government will be RMB6.75 billion (approximately US$1.10 billion on 11 December 2012, Yahoo! Finance 2012) from the Hong Kong government, RMB7 billion (approximately US$1.12 billion on 11 December 2012, Yahoo! Finance 2012) from the Guangdong-Central government, and RMB1.98 billion (approximately US$0.32 billion on 11 December 2012, Yahoo! Finance 2012) from the Macau Special Administrative Region government. The total contribution from the three governments will be RMB15.73 billion (approximately US$2.52 billion on 11 December 2012, Yahoo! Finance 2012), which will be equal to 43 per cent of the bridge's construction cost. The remainder will be financed by bank loans (Information Services Department 2008b).

The new arrangement has meant that the Guangdong government has become the largest stakeholder of the project (Hong Kong, Guangdong and central government and Macau government will take up approximately 43 per cent, 45 per cent and 13 per cent respectively of the upfront payments (Lam and Lai 2008)). In the original proposal the Hong Kong government would have taken this role. The move for this change indicates that the Chinese government is keen to push the project ahead. But there has been no comprehensive answer from the governments as to why the PPP arrangement was not adopted for the bridge (Chen and Lee 2008)

Attractive factors of the Hong Kong–Zhuhai–Macau Bridge

Maximise financial resources

To a government, PPP frees up fiscal funds for other areas of public service, and improves cash flow management as high upfront capital expenditure is replaced by periodic service payments and provides cost certainty in place of uncertain calls for asset maintenance and replacement. Public sector projects delivered via the private sector normally involve private sector funding. Consequently, the public funding required for public services can be reduced and redirected to support sectors of higher priority, such as education, healthcare and community services (Li *et al.* 2005b; Efficiency Unit 2002). Given this observation, the factor 'a. Solve the problem of public sector budget restraint' was given a relatively low score of 1, whereas the factors 'c. Reduce public money tied up in capital investment' and 'n. Non-recourse or limited recourse to public funding' were both given a score of 3.

Improve economic development

Mr Wu had observed the added advantages to industry coming from an improved infrastructure network in the Pearl River Delta region. But no further actions were taken by the Hong Kong government, and so the project was put on hold for over two decades (Oriental Newspaper 2008).

It was not until September 2002 that the project was rethought. At the Third Meeting of the Mainland/Hong Kong Conference on the Co-ordination of Major Infrastructure Projects it was agreed that a study would be conducted on transportation between Hong Kong and Pearl River West. This was the first proper study conducted to analyse the feasibility of the HKZMB. Furthermore, in January 2003, the National Development and Reform Commission (NDRC) and the Hong Kong government commissioned the Institute of Comprehensive Transportation to conduct the study, which was completed in July 2003. The report, entitled *Transport Linkage between Hong Kong and Pearl River West*, showed that transportation between Hong Kong and Pearl River West is insufficient. This point was mentioned over twenty years ago, but only now verified. The current transport between these jurisdictions via the Humen Bridge is costly and time consuming. Therefore, the report concluded that the HKZMB would be advantageous to overcome the problems (Transport and Housing Bureau 2008a). Given the evidence provided, the following attractive factors were all awarded a relatively high score of 4 for their likelihood of occurrence in this project if PPP was to be opted: 'b. Provide an integrated solution (for public infrastructure / services)'; 'j. Benefit to local economic development'; 'k. Improve buildability'; and 'l. Improve maintainability'.

Reduction in time and cost

Public sector projects delivered by the PPP model can often be completed on time and even with time savings, because the consortium would start receiving revenue once the facilities or services are up and running. Therefore, the project team is keen to complete design and construction as quickly as possible. Once it starts to accrue revenue it can begin to pay off the initial costs and build up profits, whereas in a traditionally procured project there are no extra financial incentives for public servants to deliver projects faster. As a result, projects can at least proceed as scheduled (Akintoye *et al.* 2003; Efficiency Unit 2003; Environment, Transport and Works Bureau 2004; Grimsey and Lewis 2004; Li 2003). Therefore, 'd. Cap the final service costs' was given a high score of 4, 'f. Reduce the total project cost', 'i. Reduce public sector administration costs' and 'o. Accelerate project development' were all given a score of 2 and also 'g. Save time in delivering the project' was given a score of 1.

Increase innovation

The factor 'e. Facilitate creative and innovative approaches' was also given a score of 3. Innovation is another important advantage that the private sector can bring to public services. Generally speaking, the public sector may not be as innovative as the private sector. The private sector on the other hand is continuously searching for new products and services to increase their competitive edge and to save costs (Akintoye *et al.* 2003; British Columbia 1999; Chan *et al.* 2006; Efficiency Unit 2002; Efficiency Unit 2003; Environment, Transport and Works Bureau 2004; Li *et al.* 2005b; Li 2003; New South Wales Government 2006). With regard to the local situation the attractive factor 'm. Technology transfer to local enterprise' was given a score of 2.

Risk transfer

The attractive factor 'h. Transfer risk to the private partner' was given a score of 3. Risk transfer is one of the main reasons for adopting the PPP approach. The private sector is in general more efficient in asset procurement and service delivery, and as a result it is to a government's advantage to share the associated risks with the private sector. In line with widely accepted principles, the Hong Kong government's Efficiency Unit (2003) advocated that the most ideal situation is to allocate the risk to the party most able to manage and/or control that risk. For example, the contractor would take up the construction risk, the designer would take up the design risk, the government would take up environmental approval risks, land acquisition risks, and so on (Akintoye *et al.* 2003; Boussabaine 2007; British Columbia 1999; Chan *et al.* 2006; Corbett and Smith 2006; Efficiency Unit 2002, 2003; European Commission Directorate 2003; Grimsey and Lewis 2004; Ingall 1997; Li *et al.* 2005a; Li 2003; New South Wales Government 2006; United Nations Economic Commission for Europe 2004).

Negative factors of the Hong Kong–Zhuhai–Macau Bridge

Project accountability

The Cross City Tunnel project of Sydney which was delivered by the PPP model faced many problems. Due to inaccurate traffic forecasts and the high toll prices which were applied to overcome the low traffic volume, both the consortium and the New South Wales Government were highly criticised for this project (Jean 2006). Similarly if high tolls and low usage are experienced by the HKZMB, the situation would in turn limit the co-operation between the three jurisdictions and also demean the objective of the bridge. As a result, the governments may be reluctant to deliver large infrastructures jointly again. Given evidence from previous experiences 'a. Reduce the project accountability' was given a score of 3.

Financial risks for the private sector

The governments were also aware that the private sector lacked motivation for this project. As the bridge was found to be highly costly and uneconomic, the enthusiasm of the private sector, even with compensation, would be difficult to attract (Van der Kamp 2008). Bearing these facts in mind the negative factor 'b. High risk relying on private sector' was given a score of 3.

A major reason why the West Harbour Crossing in Hong Kong was so unsuccessful compared to the Cross Harbour Tunnel, also in Hong Kong, was because it was built thirty years later and at a cost of twenty-three times more. Similarly, it has been twenty-five years since the idea for the HKZMB was first mentioned, during which time the cost of construction and the necessity for the bridge has definitely changed. The lack of interest from the private sector may be an indication that the bridge is not as important as it once was. The original intention was that the bridge could service the industrial development of the area rather than the needs of the general public. But since the idea was first proposed there has undoubtedly been a great change and movement in the region's industries. Therefore, the negative factors 'g. High participation costs' and 'h. High project costs' were both given high scores of 4.

Lack of government commitment

With the comfortable reserve from all three governments, there has been less drive to force the project as a PPP. A string of recent projects in Hong Kong has also been linked with PPP, but none of these have actually been carried through, for example the West Kowloon Cultural District. The Hong Kong government has been indecisive on the procurement methods to be used, so there is a chance that their views could also affect the Guangdong and Macau governments. For this reason, 'c. Very few schemes have actually reached the contract stage (aborted before contract)' was given a score of 3.

Lengthy delays

Lengthy delays would be one of the main concerns if the HKZMB was to be delivered by PPP. The Chief Executive of Hong Kong, Mr Donald Tsang, spoke publicly at the Eleventh Plenary of Hong Kong-Guangdong Co-operation Joint Conference in Guangzhou during August 2008 on the advantages of the HKZMB being funded jointly by the governments. He explained that the governments taking up the financing responsibility would speed up the construction work of the bridge.

This was also agreed by Chen (2008) who claimed that the project would be delivered two years earlier than the PPP approach which normally would require a lengthy consultation period and complicated legislative requirements.

The extended duration of the tendering and negotiation process due to the project being a PPP was foreseeable. Also, the differences in legislation between the three jurisdictions made it even more difficult to come up with a unique agreement on aspects such as vehicle flow and sharing of risks between the public and private sectors. As a result, continuing with the PPP plan would mean that the timeframe for the project would be more unpredictable and completion much further away (Apple Daily 2008).

The governments' decision over this project has also been supported by some of the media. If the project continued as a PPP, the private sector would need to prepare a bid based on their financial benefits in which they would take into account their expenditure for the project, the traffic forecast and the toll price. Bid preparation is a lengthy and costly process in PPP-type projects (Zhang 2001). If the governments were to find their proposals unsuitable, the process would be further extended. Similar situations to the Kai Tak Cruise Terminal in Hong Kong could arise. From the evidence available, the negative factors 'd. Lengthy delays because of political debate', 'i. A great deal of management time spent in contract transaction' and 'm. Lengthy delays in negotiation' were regarded as highly foreseeable for the HKZMB and were given the maximum score of 5 for likelihood of occurrence.

High charges for the general public

Another problem foreseeable if the project was to be delivered by the PPP model would be the high toll fees that may be imposed. The private sector comprises profit-making organisations, and so would adjust the toll fees to cover their expenditure during the delivery and maintenance of the project. In addition, they will hope to seek reasonable financial rewards. In doing so there is a risk that the project would follow in the footsteps of the East and West Harbour Crossings in Hong Kong (Apple Daily 2008). These projects were procured under the BOT model, and unlike the Cross Harbour Tunnel (Hong Kong's first and probably most successful BOT project) they suffered much bad publicity because of the high and continually increasing toll prices. As a result, the general public has tended to use the cheaper Cross Harbour Tunnel more frequently than the other two tunnels when crossing between Hong Kong Island and Kowloon Peninsula.

In the case of the HKZMB, the general public could also choose to travel on cheaper routes if the prices were to be too high. Studies showed that the HKZMB would not be commercially viable hence that would mean that the governments would have to cover the financial costs if the bridge was to be delivered by PPP (Brown 2008). If the private sector was to be involved it would be left with no choice but to raise the toll incredibly to compensate for their expenses, as in the case of the East and West Harbour Crossings (Apple Daily 2008).

Originally calculations showed that if the bridge was procured by the PPP model the toll fares would be approximately HK$150 (approximately US$19.35 on 11 December 2012, Yahoo! Finance 2012) for each vehicle crossing the bridge (Mak 2008), but whether this price will be lowered under public financing is still unknown (Hong Kong Special Administrative Region Government 2008). Chairman of the Container Transportation Employees General Union, Mr Chiu, spoke publicly that the toll fees should be lowered between the range HK$80–100 (approximately US$10.32–12.90 on 11 December 2012, Yahoo! Finance 2012) to be reasonable for the general public (Mak 2008). Another local Hong Kong car rental enterprise believed that for such a short journey the fare should not be beyond HK$100–120 (approximately US$12.90–15.48 on 11 December 2012, Yahoo! Finance 2012). The responses show that the proposed toll prices would be far too high for the general public to benefit from the project.

An analysis was conducted by So (cited in Ming Pao 2008b) on the probable toll fees under different financing models. Three different scenarios were considered. The first and second scenarios estimated the toll fee for crossing the HKZMB, with the project financed by the host governments according to a 120-year and 60-year investment return period, respectively. Other factors considered in the estimation include the savings from the private investor's profits if the project was to be financed by the PPP model, and also the estimated annual usage of the bridge. The toll fees estimated per trip were RMB$193 and RMB$387 (approximately US$30.98 and 62.12) for the first and second scenarios respectively. The third scenario considers the project under the PPP model. The investment return period was set at thirty years which is also a typical concession period for PPP-type projects (Howes and Robinson 2005). Other factors considered in the estimation also included the estimated annual usage of the bridge. Under this scenario the toll fee was calculated to peak at RMB$830 per trip (approximately US$133.23 on 11 December 2012, Yahoo! Finance 2012). Analyses of these scenarios have illustrated that the use of the PPP financial model (Scenario 3) may be two (Scenario 2) to four times (Scenario 1) more expensive than if it is funded primarily by the government. Therefore 'e. Higher charge to the direct users' was also given the maximum score of 5.

Allegation of collusion between the public and private sectors

There are also other advantages that have been perceived of the new arrangement. Chen and Lee (2008) quoted a Hong Kong academic saying that the new arrangement will minimise the chance of negotiation between developers and the governments, and hence will reduce allegation of collusion between business and the government. Hong Kong has previously been criticised for favouring certain developers and giving them high financial returns through delivering public projects. An example is the Cyberport project, a technological centre, and the West Kowloon Cultural District, a proposed cultural hub (both in Hong Kong) (Wong 2008). Given the previous experiences of Hong Kong, 'k. Confusion over government objectives and evaluation criteria' was given a score of 2 and 'l. Excessive restrictions on participation' was given a score of 3.

Other negative factors

Other negative factors include those related to staffing issues and lack of experience or skills. Although no related information was sourced for the HKZMB project, these negative factors may possibly be foreseeable. Therefore both 'f. Fewer employment positions' and 'j. Lack of experience and appropriate skills' were both given a score of 2.

Final assessment of the Hong Kong–Zhuhai–Macau Bridge

With the identification of the weightings for the attractive and negative factors of PPP, these could be identified as checklists for assessing the suitability and/or feasibility of using PPP. If the attractive factors are prevailing in a given project scenario, the use of PPP will be more positive. Conversely, if the negative factors are dominant, PPP might be considered as unsuitable. Tables 7.1 and 7.2 summarise the assessment of the HKZMB according to their calculated weightings. The findings show that the attractive factors scored between 0.06 and 0.30, whereas the negative factors scored between 0.13 and 0.45, indicating that on average the negative factors were more prevailing. The highest scoring attractive factor was found to be 'b. Provide an integrated solution (for public infrastructure/services)' and the highest scoring negative factor was 'd. Lengthy delays because of political debate'. As discussed in previous sections of this chapter, these factors are well supported with much evidence. The total scores for the attractive and negative factors in respect of the HKZMB are 2.81 and 3.58 respectively. This result shows that the negative factors are much more dominant than the attractive factors by 27 per cent, hence PPP is not the suggested procurement method for the HKZMB.

Additional attractive and negative factors

The attractive and negative factors discussed in this chapter were derived from a comprehensive literature review and verified by previous researchers. Their relevance and appropriateness was also confirmed from data collected around the world. The survey respondents were also given the opportunity to suggest additional factors to ensure that the ones already derived were representative for PPP projects. Nevertheless, the questionnaire surveys sought opinions on PPP projects in general and not for particular projects. Therefore it is still anticipated that there may be additional factors depending on the case being considered due to the uniqueness of each project. As a result of the HKZMB analyses, there are a couple of additional factors which should also be considered if a proper evaluation model for assessing the suitability of PPP for this case was to be conducted.

From the case analyses of the HKZMB, the attractive factor 'Uplift public image' should also be considered. If the public image of a project can be uplifted as a result of being delivered by PPP then this should be an additional attractive factor which needs to be considered. The analyses demonstrated how the general public's opinion was also vital in reflecting a project's success. In the case of the HKZMB its public image had faded immensely because of the lengthy discussion on financing options. The support from the general public is vital as they also represent the future end-users of the facilities and services.

One recent argument over PPP projects in Hong Kong is whether the private sector is given too much financial benefit in return for providing the services and

Table 7.1 Mean scores and rankings for the attractive factors of PPP

Attractive factors	N	Mean	Rank	Weighting	Assessment of the HKZMB	
					Likert scale assessment	Score
a. Solve the problem of public sector budget restraint	34	3.65	3	7.16%	1	0.07
b. Provide an integrated solution (for public infrastructure/services)	33	3.79	1	7.43%	4	0.30
c. Reduce public money tied up in capital investment	33	3.48	6	6.82%	3	0.20
d. Cap the final service costs	34	3.26	10	6.39%	4	0.26
e. Facilitate creative and innovative approaches	34	3.74	2	7.33%	3	0.22
f. Reduce the total project cost	33	3.09	14	6.06%	2	0.12
g. Save time in delivering the project	34	3.21	13	6.29%	1	0.06
h. Transfer risk to the private partner	34	3.65	4	7.16%	3	0.21
i. Reduce public sector administration costs	33	3.39	8	6.65%	2	0.13
j. Benefit to local economic development	34	3.56	5	6.98%	4	0.28
k. Improve buildability	33	3.24	11	6.35%	4	0.25
l. Improve maintainability	34	3.32	9	6.51%	4	0.26
m. Technology transfer to local enterprise	34	2.94	15	5.76%	2	0.12
n. Non-recourse or limited recourse to public funding	34	3.21	12	6.29%	3	0.19
o. Accelerate project development	34	3.47	7	6.80%	2	0.14
Total	51			100.00%	42	2.81

Source: Cheung and Chan 2011a (with permission from the American Society of Civil Engineers).

Table 7.2 Mean scores and rankings for the negative factors of PPP

Negative factors	N	Mean	Rank	Weighting	Assessment of the HKZMB	
					Likert scale assessment	Score
a. Reduce the project accountability	34	2.79	12	6.60%	3	0.20
b. High risk relying on private sector	34	3.09	10	7.31%	3	0.22
c. Very few schemes have actually reached the contract stage (aborted before contract)	34	3.41	3	8.07%	3	0.24
d. Lengthy delays because of political debate	34	3.82	1	9.04%	5	0.45
e. Higher charge to the direct users	34	3.26	9	7.72%	5	0.39
f. Fewer employment positions	34	2.79	13	6.60%	2	0.13
g. High participation costs	34	3.35	5	7.93%	4	0.32
h. High project costs	34	3.03	11	7.17%	4	0.29
i. A great deal of management time spent in contract transaction	34	3.29	6	7.79%	5	0.39
j. Lack of experience and appropriate skills	33	3.27	8	7.74%	2	0.15
k. Confusion over government objectives and evaluation criteria	34	3.41	4	8.07%	2	0.16
l. Excessive restrictions on participation	34	3.29	7	7.79%	3	0.23
m. Lengthy delays in negotiation	33	3.45	2	8.17%	5	0.41
Total		42.25		100.00%	46	3.58

Source: Cheung and Chan 2011a (with permission from the American Society of Civil Engineers).

facilities. The media have often suggested the existence of public–private collusion. The HKZMB did not attract much private interest; providing a more attractive business case would not have been evitable. But due to the continuous hyped-up critique of providing private sectors with unreasonably high financial benefits, the government has been careful to avoid such allegations. In this case, the government was able to avoid these as they paid for the project from the public purse. Therefore, the negative factor 'Suspected public–private collusion' should also be included as one of the negative factors.

The HKZMB case study has shown that at different times and stages, the attractive and negative factors may vary slightly depending on the project it is applied to and its geographic location.

Chapter summary

This chapter presents an evaluation model for evaluating the suitability of PPP projects. Using this model, potential PPP projects can be assessed and assigned a score for their attractive and negative factors. The HKZMB was used to demonstrate the feasibility of this model. The results show that the negative factors outweigh the attractive factors of this project hence the use of PPP to deliver this project would not be recommended. This suggestion also falls in line with the actual decision made by the host governments of this project. This evaluation model has presented a system for users to analyse whether potential public projects should be procured by PPP. The method presented is believed to be useful for both the public and private sectors especially during the early stages of project evaluation.

8 Risk factors of public–private partnerships

Introduction

With the growing economic development experienced around the globe, there is an urge for more and better public infrastructure. Public–private partnership (PPP) is an innovative method of delivering these facilities and services. But along with this method come certain risks that exist or are more severe when compared to the traditional delivery method. A study was conducted by Cheung and Chan (2011b) looking at three categories of common public projects in China that are often delivered by the PPP method, including Water and wastewater, Power and energy and Transportation. China is one of the countries where a new breed of PPP projects has started to evolve. Although the study presented in this chapter focuses heavily on China, it is believed that the principles behind these findings are timely and equally applicable to other countries and jurisdictions. In this study, for each type of project, experienced practitioners in China were asked to rank the severity of twenty risk factors sought from a comprehensive literature review. The top five most severe risk factors for each type of project were considered. Ranked severe for all three types of projects were 'Government intervention' and 'Public credit'. The findings indicate that the most severe risks are government-related. It appears that the stakeholders have low confidence in the government. These findings have highlighted the severity of risk factors for common types of PPP projects in China. With this information both the public and private parties can be more aware of which risk factors would be the most severe for certain projects. As a result, appropriate precautions can be made to avoid or minimise the likelihood and consequences of these risks. By doing so PPP projects can be carried out more successfully and their further use can be encouraged around the globe. PPP stakeholders from other countries can also use the findings presented in this study to prevent potential risks from occurring.

Risk factors of PPP projects

Projects procured by PPP tend to be subject to more risks compared to those projects that are procured traditionally, because of their complexity. Ke *et al.*'s (2009) study confirmed that risk management (including risk identification, risk evaluation, risk allocation, risk management, financial risk, political risk and market risk) has continued to be one of the main research interests of PPP in recent years. Furthermore, Khasnabis *et al.* (2010) emphasised the importance for future PPP studies to consider risk and uncertainties. Consequently, this section explores some of the studies which have been conducted by previous researchers in the area of PPP risks.

Unkovski and Pienaar (2009) considered the management and analysis of PPP risks. Their results showed that although there were many risks associated with PPP projects, the method is still considered to be advantageous in South Africa because they are lower in cost and more manageable when compared to using the traditional method whereby the government finances and delivers the project itself. Three major types of risks were categorised in their study, namely technical, financial and legal risks.

Chen and Shi (2009) identified similar risks for PPP projects but also provided a different perspective on how they should be considered. They defined PPP risks according to two main groups: systematic risks and non-systematic risks. Systematic risks refer to those that are caused externally and cannot be controlled by the concessionaire. They include political risk, legal risk, financial risk and contingent risk. On the other hand, non-systematic risks are those which are related to the project construction and operation. These can include completion risk, operation risk and market risk.

The Efficiency Unit (2008) of the Hong Kong Special Administrative Region in China classified in their guideline for conducting PPP projects the key types of risks; these include: demand risk; design and construction risks; operation and maintenance risks; technology/obsolescence risk; finance risk; legislative risk; approval risk; and hazard risk. Again, similar risks were identified by different researchers irrespective of geographical location.

Chen and Doloi (2008) conducted a comprehensive literature review looking at the factors holding back PPP projects in China and generally. They found that those factors specific to China include: opaque and weak legal systems; complex approval systems; regulatory constraints on market entry; low market prices for infrastructure products and services; creditworthiness of local utilities; no direct interests to local government and its subordinates; and foreign currency administration difficulty. Unique about these PPP risks that were identified for China is that they are all related or affected by the local government in one way or another. Previous research may indicate that the government should take more responsibility for providing a suitable environment to engage PPP projects.

Li and Zou (2011) derived slightly different findings from their study. They presented a fuzzy analytical hierarchy process as a risk assessment technique for a PPP expressway project in China. The results showed that planning deficiency, low project residual value at the end of the concession period, lack of qualified bidders, design deficiency and long project approval time were assessed as the top five risks for the project.

Furthermore, Li and Liu (2009) suggested that to implement PPP projects in China, the risks of the project needs to be considered from different angles, including: in terms of curiosity; as long-term, complex and multi-level risks; and also bearing in mind the multi-goals of stakeholders. They firmly believe that the severity of the risks would differ depending on whether it is a traditional or PPP project being considered.

The effective handling of risks is often related to the appropriate risk allocation between the public and private sectors. Ke *et al.*'s (2010) study aimed to identify the preferred risk allocation of PPP risks in mainland China, Hong Kong, the UK and Greece. The results of their study indicated that political, legal and social risks should be handled by the public sector in mainland China and Hong Kong. Other researchers have also demonstrated different techniques for handling risk allocation. For example, Jin (2011) found that neuro-fuzzy models could be used to forecast efficient risk allocation strategies for PPP projects at a highly accurate level, which would be impossible using multiple linear regression models and fuzzy inference systems. The same researcher conducted a previous study (Jin 2010) which considered the features related to risk allocation in PPP projects, including partners' risk management routine, partners' risk management mechanism, partners' co-operation history, risk management environmental uncertainty, and partners' risk management commitment. These features were used as determinants in the decision-making process of efficient risk allocation.

Duffield (2001) took another step forward to propose a risk evaluation technique to assess the severity of risks for different PPP projects. The likelihood and consequence of the risk would be represented by a risk index. Furthermore, the risk index would be defined according to four categories of severity which would suggest the approach for handling the risk. These categories include: (1) Rely on procedures and contract administration to manage risk; (2) Line management awareness and control; (3) Director awareness; and (4) Ministerial awareness. Similarly, Pantelias and Zhang (2010) proposed a methodological framework to evaluate the financial risk of transportation infrastructure projects delivered by PPP. They claim that the approach is simple to use and effective for considering investment options through scenario and sensitivity analyses.

Research approach

Data for Cheung and Chan's (2011b) study were mainly collected via interviews conducted with experienced practitioners in China. The respondents were asked to rank the importance of risk factors for the three types of projects: Water and wastewater, Power and energy and Transportation. Ranking and prioritisation of risks in PPP projects is an important part of risk management, ensuring that risks can be effectively allocated to the most appropriate party (Iyer and Sagheer 2010). This section describes the design of the interview template and the background of the interview respondents. Furthermore, the analytical techniques adopted are explained. These include: mean score ranking, Cronbach's alpha and Kendall's concordance analysis.

Design of interview template

In order to analyse the risk ranking and allocation for different types of PPP projects in China an interview template was designed and conducted with PPP experts. Respondents were asked to provide some simple background information related to their experience. They were also presented with a list of twenty PPP risk factors and asked to rate them according to their severity according to a Likert scale from 1 to 5, with 1 representing the least severe and 5 representing the most severe. The list of risk factors was derived based on a comprehensive literature review and also from findings of a previous questionnaire survey conducted by the authors and their research team (Xu *et al.* 2010). To prevent misinterpretation, the interview respondents were provided with the definition for each of the twenty PPP risk factors as shown in Table 8.1.

Background of interview respondents

A total of thirty-eight interviews were conducted in major cities around China including Beijing, Shanghai, Nanjing and Dalian. The selection of these cities was based on their rapid development in infrastructure, their activeness in PPP projects, and also the available contact points of the researchers. General information regarding the respondents' background was recorded, including the number of years they have been involved with PPP projects, the number of PPP projects they have participated in, the type of sector they were working for, and also the types of projects that they have been involved with. All respondents participated in the interviews have hands-on experience with PPP projects (Figure 8.1). The majority (63 per cent) have five years' or less working experience. Approximately 30 per cent of the respondents had six to ten years of working experience and the remaining had more years. This experience profile is considered acceptable given that PPP projects have only become more popular in China in recent years. All respondents interviewed were experienced in running PPP projects (Figure 8.2). All respondents had executed at least one PPP project; 66 per cent of the respondents had executed one to three projects, a few had executed four to five projects and approximately 10 per cent had executed six or more. A large proportion of the respondents (43 per cent) represented the private sector, 34 per cent represented other organisations, and fewer respondents represented the public sector (Figure 8.3).

Seven types of projects that the respondents have been involved with were identified (Figure 8.4). In order of highest involvement, these included fifteen in Water and wastewater (39 per cent), eight in Power and energy (21 per cent), seven in Transportation (18 per cent), four in Other types of project (11 per cent), two in Housing and offices (5 per cent), and also one in Hospitals and medical services (3 per cent) and Cultural and sport facilities (3 per cent) respectively. These projects represented proportions as shown in the brackets. Considering that the first three types of projects were dominant in terms of participation level, they were selected for comparison purposes in this study based on the severity of their risk factors.

Table 8.1 Definition of PPP risk factors

Risk no.	Risk	Definition
1	Government intervention	Public sector interferes unreasonably in the facilities/services
2	Public credit	The reliability and creditworthiness of the government to fulfil obligations
3	Financing risk	Financial difficulties experienced by the consortium as a result of poor financial market or lack of financial income
4	Poor public decision-making process	Government makes wrong or poor decisions due to lack of knowledge or interest
5	Subjective project evaluation method	Subjective evaluation at the beginning of a public project to decide the procurement method
6	Completion risk	Project takes longer than the predicted time to complete
7	Government corruption	Bribery of bureaucrats resulting in inappropriate privileges and benefits being offered to the private sector
8	Price change	Improper tariff design or inflexible adjustment framework leading to insufficient income
9	Operation cost overrun	Operation cost overrun resulting from over-priced operation and slow operation
10	Imperfect law and supervision system	Lack of specific laws for PPP projects
11	Project/operation changes	The likelihood of unexpected changes and errors occurring during the project operation
12	Inability of concessionaire	The consortium not being able to perform its obligations as agreed
13	Inflation	Unanticipated changes to inflation rate
14	Conflicting or imperfect contract	Improper arrangements in the contract such as inappropriate risk allocation amongst stakeholders
15	Interest rate fluctuation	Unanticipated fluctuations to interest rate
16	Insufficient project finance supervision	The financial status and expenditures are not monitored and controlled
17	Delay in project approvals and permits	Delay or refusal of project approval or permit by government
18	Inadequate competition for tender	Lack of transparency and structure during tender, lack of opportunities for tenderers, few tenderers
19	Foreign exchange fluctuation	Fluctuation in currency exchange rate and/or conversion difficulties
20	Change in market demand (non-competition factor caused)	Demand change, the need for the services and facilities have changed, maybe not needed or less needed than before

Source: Cheung and Chan 2011b (with permission from the American Society of Civil Engineers).

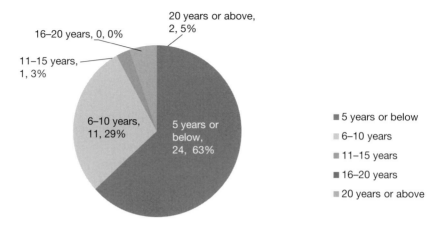

Figure 8.1 Years of experience in working with PPP projects for the respondents

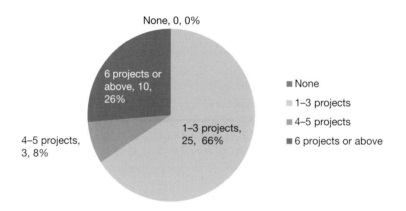

Figure 8.2 Number of PPP projects the respondents have participated in

Reliability and agreement of survey results

The value of Cronbach's alpha was calculated to be 0.822 indicating that a high level of uniformity amongst the survey responses was received (Norusis 2008).

The Kendall's coefficient of concordance (W) for the ranking of risk factors was 0.406. The computed W was statistically significant with significance level at 0.000.

As the number of attributes considered was above seven, as mentioned previously the chi-square value would be referred to rather than the W value. According to the degree of freedom the critical value of chi-square was 30.144. The computed chi-square value was found to be above this value at 115.852. Therefore, the

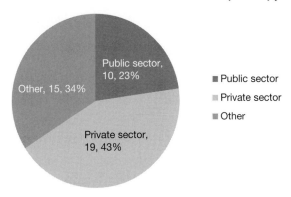

Figure 8.3   Sector to which the respondents belong

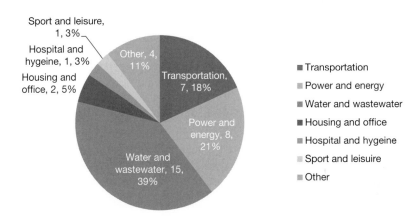

Figure 8.4   Types of projects which the respondents have been involved with

assessment by the survey respondents on their rankings of risk factors is proved to be consistent. This finding ensures that the completed survey questionnaires are valid for analysis.

Ranking of risk factors

The twenty risk factors were rated by interviewees according to the severity of their threat towards different types of PPP projects (Table 8.2). The mean score ranking technique was used to analyse the results obtained from these interviews. The interviewees were asked to assess the risks according to a Likert scale from 1 to 5 (1 = least important and 5 = most important). The mean score for each risk was therefore calculated by the summation of the respective scores given by each respondent according to the Likert scale, divided by the number of respondents who assessed the risk. The results were then further ranked and studied for Water and wastewater, Power and energy, and Transportation projects.

The ranking of the top five most severe risk factors for each type of project was identified and analysed. In total nine risk factors were studied. The following discussion aims to provide some reasons why these risk factors are believed to be the most severe. In addition, the risk factors of the three types of projects were compared to identify similarities and differences.

Government intervention

The risk factor 'Government intervention' was ranked in the top five amongst the twenty risk factors for all three types of projects. For Power and energy and Transportation projects this risk was ranked the most severe. For Water and wastewater projects this risk was ranked slightly lower at fourth position. Qi *et al.* (2009) conducted an analysis of sixteen PPP projects in China. These projects included those from the Water and wastewater, Power and energy, and Transportation sectors. From their analyses 'Government intervention' was a primary cause of failure recorded. Government intervention would only be appropriate if without it the general public would be substantially affected. For example, if unacceptably high toll fees or service fees are charged to the general public, the government would probably consider stepping in to restrict the consortia. Obviously, government intervention would only be feasible if it is also contractually viable. Otherwise, unreasonable government intervention would ruin the relationship with the private sector and discourage their interest in future PPP projects. Zhong and Fu (2010) also reported that some of the early PPP projects in Guangdong failed because they were implemented solely by the local government without professional advisers, showing a high level of government intervention.

Public credit

Also ranked in the top five for all three types of projects was 'Public credit'. Transportation projects were ranked slightly higher at second position; whereas Water and wastewater and Power and energy projects were both ranked fifth place. The findings are in line with Sachs *et al.*'s (2007) discussion regarding the creditworthiness of the local governments in China. They highlighted that one of the main problems related to the application of PPP in China was the unrealistic and unreasonable guarantees made by Chinese local governments. As a result, public credit has become a concern. They further discussed that the Chinese local governments usually make promises which they are incapable of keeping in order to attract potential investors to carry out the projects. Unfortunately, contracts are frequently breached because of this common practice of the Chinese local governments. As a result, both parties lose out. The private party may lose their investment or achieve unexpectedly lower returns than anticipated and achieve no compensation. Sachs *et al.* (2007) concluded that the Chinese local governments have been known to pay more in order to resolve the damage which has been caused to the other parties or the project itself.

Table 8.2 Comparison of risk ranking amongst different project sectors

Risk no.	Name of risk factor	Types of project					
		Water and wastewater		Power and energy		Transportation	
		Ranking	Mean	Ranking	Mean	Ranking	Mean
1	Government intervention	4.14	4	3.98	1	4.00	1
2	Public credit	4.00	5	3.70	5	3.91	2
3	Financing risk	4.71	1	3.16	9	3.12	8
4	Poor public decision-making process	4.00	6	3.33	7	3.49	5
5	Subjective project evaluation method	4.33	3	3.87	3	3.24	7
6	Completion risk	4.43	2	2.59	14	2.85	16
7	Government corruption	3.17	12	3.87	2	2.98	10
8	Price change	3.25	9	3.06	10	3.81	3
9	Operation cost overrun	3.29	8	3.64	6	3.05	9
10	Imperfect law and supervision system	3.00	13	3.31	8	3.61	4
11	Project/operation changes	2.83	14	2.12	18	3.35	6
12	Inability of concessionaire	2.60	16	3.81	4	2.96	11
13	Inflation	3.33	7	2.50	15	2.53	20
14	Conflicting or imperfect contract	3.20	10	2.36	16	2.87	14
15	Interest rate fluctuation	3.20	11	2.61	11	2.69	18
16	Insufficient project finance supervision	2.75	15	2.60	13	2.88	12
17	Delay in project approvals and permits	2.57	=17	2.10	19	2.85	15
18	Inadequate competition for tender	2.57	=17	1.82	20	2.80	17
19	Foreign exchange fluctuation	2.57	=17	2.33	17	2.66	19
20	Change in market demand (non-competition factor caused)	1.88	20	2.61	12	2.88	13

Source: Cheung and Chan 2011b (with permission from the American Society of Civil Engineers).

Financing risk

'Financing risk' has always been a major problem especially for Water and wastewater projects. For example, the Guangzhou Xilang project which was the first PPP wastewater treatment plant project in China was held back because of financing risk. It was initially planned in 1993 but took several years to take off because of the lack of financing source (Zhong and Fu 2010). Another example occurred in 2004, where the Beijing government introduced five small-sized wastewater treatment plant projects. These projects aimed to improve the wastewater treatment capacity and control water pollution in Beijing. Unfortunately the awarded consortia withdrew from the project because of financial difficulties. Chinese banks are often reluctant to provide the long-term loans which are required for PPP projects, or tend to restrict the credit policies to the private sector. These experiences have reflected the problems in the existing financing policies of China (Zhong and Fu 2010). Consequently, 'Financing risk' was ranked the most severe amongst the twenty risk factors for Water and wastewater projects. For the other types of projects studied, this risk was ranked of medium severity only. It must also be noted that the financing model adopted for each project will vary its level of financial risk. This study focuses primarily on comparing different-natured PPP projects only. For further studies it would be worthwhile for researchers to consider how financing risk is affected by the mode of PPP adopted in projects.

Poor public decision-making process

The risk factor 'Poor public decision-making process' was ranked similarly for the three types of projects. Transportation projects were ranked slightly higher at fifth position, possibly indicating that the Chinese government is more prone to making poor decisions for these types of projects. In Sachs *et al.*'s (2007) study they reported that wrong decision making by the Chinese government was another problem holding back the implementation of PPP. This was ascribed to the lack of knowledge in running PPP projects and also the unrealistic guarantees which were made by the Chinese government. As a result, there has been much complaint from the general public, and key officials have stepped down (Sachs *et al.* 2007).

Subjective project evaluation method

'Subjective project evaluation method' was ranked third for Water and wastewater and Power and energy projects but only of medium severity for Transportation projects. The reason behind this difference is probably the fact that traditionally Water and wastewater and Power and energy projects have been handled by the government. But since the 1990s the Chinese government has started to introduce private financing for these projects (Zhong *et al.* 2008). With private financing as the target, proper evaluation of projects has been neglected. An all-round evaluation should be conducted in order to assess whether PPP would be the suitable method for delivering certain public projects. The evaluation criteria should focus on value for money, innovation, expertise, time, cost, general public satisfaction, and so on. Khasnabis *et al.* (2010) stressed the importance of conducting a careful analysis before PPP projects are undertaken to assess the financial and economic implications of the project from each participant's viewpoint, with due regard to risks and uncertainties associated with such long-term investments. Unfortunately private financing has been a priority for adopting the PPP approach for those ex-government-run projects. Consequently, all-round evaluations of the projects have not been conducted adequately. The interviewees reflected in their ranking the importance of a subjective project evaluation method.

Completion risk

'Completion risk' was only ranked highly for Water and wastewater projects, at the second position. Generally speaking completion risk causes a project to go beyond the initial schedule. The consequences are a lack of cash flow to pay for the operating costs and subsequent debts, postponed length of maturity, and increased interest from the loan (Li and Liu 2009). As a result the whole project cost will be increased and the project will not be completed as planned. Furthermore, Pribadi and Pangeran (2007) analysed the risks that were associated with water PPP projects. Their study found that delay in completion for water PPP projects was often caused by lack of co-ordination of contractors, failure to obtain standard planning approvals, or failure to grant contractual land use rights or rights of way. These causes probably help to explain why 'Completion risk' was ranked high for Water and wastewater projects.

Government corruption

This risk factor was regarded as a potential threat for Power and energy projects by the interviewees at the second rank. In contrast, for the other types of projects this risk factor was not regarded as threatening. Although there is no evidence to explain this large difference in ranking between the projects, government corruption has previously been suspected for Power and energy projects. The Laibin B power project was an example of successful PPP implementation and was adopted as a role model for similar projects (Sachs *et al.* 2007). In addition, it was revolutionary at the time both for being awarded through international tendering, and also for comprising 100 per cent foreign ownership. Wang and Ke (2009) believe that although the Chinese government had addressed the risk of government corruption via warranties in this project, there was no confidence that the private party could walk away easily if it did occur. They further discussed that their beliefs are due to several predictions: that corruption would not take place in the open; it is difficult to determine corruption using contract language; and also the enforcement of the contract terms would be in doubt.

Imperfect law and supervision system

The risk factor 'Imperfect law and supervision system' was ranked fourth for Transportation projects. For Power and energy projects it was ranked of medium severity. And for Water and wastewater projects it was ranked low. In many Chinese PPP projects, it is not uncommon to find that the financiers undertake roles on both sides of the PPP arrangement and often they will also supervise the project as well. The effectiveness of this arrangement can be dubious. Aware of the potential problems, some projects, especially those concerning Transportation, have taken action to avoid overlapping roles. In the Guangzhou No. 2 Underground Line project, the supervision of all aspects related to the project were purposely passed to the public procuratorial service and the financing bank acted as a double check (Adams *et al.* 2006). Other measures which have been taken to improve legislation related to Transportation projects include the establishment of specific laws such as the Highway Law (Chen and Doloi 2008). It is obvious that these actions arise because of the riskiness of an imperfect law and supervision system for Transportation projects. In some situations it is possible that the laws are simply not enforced.

Inability of concessionaire

This risk factor was ranked fourth for Power and energy projects. For the other types of projects this risk factor was ranked relatively low. Previous studies (Braadbaart *et al.* 2008; Zhong and Fu 2010) have shown that the lack of competition during the bidding process in PPP projects has meant that the wrong or inappropriate concessionaires have been selected. The result is that incapable concessionaires have been selected. It is possible that Power and energy projects are technically more demanding than the other types of projects studied, hence the ability of the concessionaire would be more crucial.

Chapter summary

This chapter has examined some of the most severe risk factors that could occur in PPP projects. A comparison was conducted looking at the risk factors of Water and wastewater, Power and energy, and Transportation projects in China. The results showed that 'Government intervention', 'Public credit', 'Financing risk', 'Poor public decision-making process', 'Subjective project evaluation method', 'Completion risk', 'Government corruption', 'Imperfect law and supervision system', and 'Inability of concessionaire' were the most severe risk factors for these projects, with 'Government intervention' and 'Public credit' being severe for all three groups of projects. It appears that the major risks of PPP projects in China are mainly related to the government. Some of the lessons learnt and recommendations from these findings include:

- The consortium members should consist of non-government representatives to avoid government intervention.
- The Chinese government should make realistic promises that they intend to and are able to keep.
- Ensure a stable income to eliminate financing risk. Ideally the income should result from the services and facilities but if this is not feasible, government

support should be considered. Special attention on this aspect should be given to Water and wastewater projects.
- The Chinese government should understand the PPP process well and try to adopt the good practices of other countries where possible.
- Currently, there is no equivalent of the public sector comparator in China to assess whether the public projects are suitable to be delivered by PPP. It is important to introduce such a process to ensure that projects are not wrongly delivered by PPP.
- Future Water and wastewater projects in particular should consider implementing an early and structured plan to avoid completion risk.
- Avoid government corruption especially for Power and energy projects. The government should enforce prosecution to eliminate the occurrence of corruption.
- Ensure that laws are enforced especially for Transportation projects.
- Concessionaires for Power and energy projects in particular should be selected carefully and appropriately to avoid under-qualified members.

This chapter has provided an interesting perspective on procuring PPP projects especially for those practitioners and academics in Western countries. The study has highlighted those most common types of PPP projects in China and analysed the differences between their risks. It is hoped that the results have enabled project stakeholders from other countries to be more aware of the potential risks in order to avoid or minimise them effectively. Furthermore, opportunities for conducting PPP projects in China will as a result be encouraged.

Part III

Public–private partnership case studies

9 Innovative social PPP projects

Introduction

This chapter presents a new innovative form of social public–private partnership (PPP) introduced by the Hong Kong government, where service providers are invited to revitalise historic buildings for new functions. Under this scheme the government will provide all the initial financial assistance required, unlike traditional PPP projects. Two case studies adopting this method are presented including the Mei Ho House (Cheung and Chan 2012) and the North Kowloon Magistracy. The purpose of this chapter is therefore to evaluate the success of this innovative scheme via these case studies.

The first case study presented is the Mei Ho House which is a Grade I listed building marking the history of early public housing in Hong Kong. This project was studied thoroughly by looking at the historical background, the selected service provider, project details, historical significance, social benefits and public opinion. The second case study, the Savannah College of Art and Design Hong Kong, formerly the North Kowloon Magistracy, was the first revitalised project under this scheme. Focus group meetings were conducted and the participants were asked to share their views regarding the effectiveness of the scheme, the impacts of the scheme, and also their views regarding general public participation.

The revitalisation scheme presented in this chapter has demonstrated how projects can benefit from the private sector's expertise but at the same time be funded by the government. This model is particularly attractive for those governments where money is not their greatest concern, but where the priority is to tap into the expertise from the private sector. It appears that positive support has been shown towards this scheme, indicating that there is a strong possibility for further developments both in Hong Kong and other similar jurisdictions.

Revitalising Historic Buildings through Partnership Scheme

Similar to other major cities around the world, economic development in Hong Kong has been criticised as running at the expense of the historical, social and cultural elements of the city. Therefore, the conservation of these elements has become increasingly important. In addition, creating a sustainable business environment for social enterprises to take care of disadvantaged groups is seen as the responsibility of government. Consequently, there is much potential for reusing historic buildings, renovated at government expense, for subsequent use by social enterprises. However, accountability is always an issue in the deployment of public funding. If public funding is not deployed properly, it could easily lead to public criticism.

The Hong Kong government introduced a Revitalising Historic Buildings through Partnership Scheme in October 2007. In this scheme the Hong Kong government will pay all initial costs related to the renovation of these buildings for subsequent renting to service providers of social enterprises and will not expect the service providers to shoulder these costs. The facilities have also been rented to the service providers at a nominal or heavily subsidised cost.

The main objectives of the scheme are: (1) to preserve and put historic buildings into good and innovative use; (2) to transform historic buildings into unique cultural landmarks; (3) to promote active public participation in the conservation of historic buildings; and (4) to create job opportunities particularly at the district level (Development Bureau 2012). It is hoped that pumping in public money to upgrade these historic buildings will in turn generate jobs, uphold conservation principles, and also subsidise the social enterprises in running their businesses which may not remain feasible without financial support. The service providers of this scheme must be non-profit-making organisations with prior experience in the area proposed. Service providers are invited to submit proposals for using these buildings to provide services or businesses in the form of social enterprises. In their submissions, they should propose detailed plans to demonstrate how these buildings would be preserved, how their historical significance can be effectively utilised and also how the enterprises would operate to show financial viability and benefit to the community. The successful service providers will be awarded a one-off financial package to support the renovation of the buildings, paying only nominal rent for the buildings and also be given a one-off grant for the initial costs. The maintenance works will be conducted solely by the service providers themselves.

The Hong Kong government has identified a total of fourteen historic buildings that they feel are appropriate for this scheme to be preserved in three batches. In the first batch seven buildings were identified as Lui Seng Chung, Lai Chi Kok Hospital, North Kowloon Magistracy, Old Tai O Police Station, Fong Yuen Study Hall, Mei Ho House and Old Tai Po Police Station. Service providers for the first six of these were selected. The unselected project Old Tai Po Police Station was re-launched in the second batch of this scheme with four other historic buildings comprising Blue House Cluster, Former Fanling Magistracy, Old House at Wong Uk Village and Stone Houses. The service providers for only three of these projects were selected including Old Tai Po Police Station, Blue House Cluster and Stone Houses. The Former Fanling Magistracy was re-launched in the third batch of historic buildings together with three other newly selected buildings which were King Yin Lei, Har Paw Mansion and Bridges Street Market. The proposals for these projects were still being considered in 2013 and the selected service providers are yet to be identified.

Although the intention of this revitalisation scheme should be praised, the Hong Kong government has still received much criticism over the appropriateness of the service providers selected. For example, the Chinese Artists Association had proposed using the North Kowloon Magistracy as a centre for Chinese opera. But their bid was unsuccessful and instead an American service provider was selected. According to the Legislative Council papers, the Savannah College of Art and Design was selected for a number of reasons. They require no government funding, they are experienced in the preservation of historic buildings, and lastly, synergy with other revitalisation projects in the district will be created (Sing Tao Daily 2009). Despite this rationale, the appropriateness of the selected service provider is doubted, as many feel that local organisations should be given higher priority.

The Mei Ho House case study

Historical background

The Mei Ho House (block 41) is part of the Shek Kip Mei Estate located in Sham Shui Po of the Kowloon Peninsula in Hong Kong. The house was an early type of public housing constructed by the Hong Kong government back in 1954 and was in full use until 2004. It was originally constructed to provide public housing for those 58,000 people who were made homeless by the fire which affected the Shek Kip Mei squatter area in 1953 (Hong Kong Housing Authority 2010). The estate consisted of twenty-nine blocks constructed using reinforced concrete frame and floors and concrete block walls and partitions (Commissioner for Heritage's Office 2010a). Mei Ho House was one of the first eight 6-storey H-shaped plan blocks and covered a gross floor area of 6,750 square metres (Commissioner for Heritage's Office 2010b). These blocks featured two wings connected by a bar and were named 'H' because of their aerial view which resembled the letter H (Hong Kong Housing Authority 2010) (Figure 9.1; Flickr 2010). The buildings each contained 384 units, each with accommodation for five adults. The rent was HK$10 (approximately US$1.3 on 11 December 2012, Yahoo! Finance 2012) per month plus another HK$1 (approximately US$0.1 on 11 December 2012, Yahoo! Finance 2012) per month for water supplies. The buildings were designed to provide basic necessities only, including two water sources and six toilets on each floor. Residents cooked in the communal corridors. Staircases were situated at the end of the buildings and there were no lifts. As part of a redevelopment plan in the 1970s, the units were converted into larger units containing toilet and kitchen facilities. With continuous redevelopment to provide more and better housing at the beginning of this century, Mei Ho House is now the only surviving H-shaped block not to be demolished, in order to conserve this important piece of historical evidence marking the beginning of public housing in Hong Kong (Commissioner for Heritage's Office 2010c). Subsequently, the house has been classified by the Antiquities and Monuments Office in Hong Kong as a Grade I building, which represents the highest possible grade amongst three, and indicates the buildings are of outstanding merit and for which every preservation effort should be made (Antiquities and Monuments Office 2010).

Figure 9.1   Aerial view of Mei Ho House (Flickr 2010) (with permission from 302 user of Flickr)

The selected service provider

In February 2009 it was announced that the Hong Kong Youth Hostels Association (HKYHA) was selected by the Advisory Committee on Revitalisation of Historic Buildings (ACRHB) as the service provider for the Mei Ho House (Commissioner for Heritage's Office 2010b; 2010d). The HKYHA was established in 1973. It currently runs seven hostels in Hong Kong and is affiliated to the International Youth Hostel Federation (Commissioner for Heritage's Office 2010b). The HKYHA was selected as a suitable service provider for this project due to its strong worldwide network, proven track record in hostel operations and managing large-scale projects, and also care for the under-privileged (Hong Kong Youth Hostels Association 2010).

The HKYHA has defined the following project goals (Hong Kong Youth Hostels Association 2010):

- To bring out the value of Mei Ho House;
- To revitalise and upkeep the building through adaptive reuse as a city hostel;
- To ensure sustainable operations;
- To contribute to the benefits of the Sham Shui Po district both socially and economically.

A working group within the HKYHA has been set up to oversee the Mei Ho House project. The following tasks will need to be completed (Commissioner for Heritage's Office 2010b):

- Complete the detailed project design;
- Complete the required administrative procedures;
- Seek planning permission from the Town Planning Board on the intended usage;
- Seek formal funding approval from the Finance Committee of the Legislative Council;
- Carry out the renovation work (anticipated to be around 18 months);
- Apply for licences for intended operations;
- Sign agreements with government on operations of the social enterprise and preservation of the historic building.

Project details

Mei Ho House will be converted into a hostel and museum by the HKYHA. The capital cost for this project will be around HK$192.3 million (approximately US$24.8 million on 11 December 2012, Yahoo! Finance 2012). The Hong Kong government will subsidise around $4.4 million (approximately US$0.6 million on 11 December 2012, Yahoo! Finance 2012) of this for the operation. The project was scheduled to take approximately 18 months and be completed in 2012 (Commissioner for Heritage's Office 2010b) but to date (May 2013) has yet to be completed.

The majority of Mei Ho House will be converted into a hostel with a section reserved as a museum marking the history of Hong Kong's public housing. The hostel will provide 109 twin or double bedrooms, nine dormitories, two family rooms and four rooms for people with disabilities. The connecting buildings in the middle will be re-constructed to contain lifts as well as common areas such as meeting rooms and dining facilities for the hostel. In addition, the rear court-yard will provide an open area for entertainment performances and functions (Commissioner for Heritage's Office 2010b).

Structural and decorative renovations will also be required for the Mei Ho House. The open corridors will be restored and veranda floor slabs will be re-cast and supported by new structural arms (Hong Kong Youth Hostels Association 2010).

Historical significance

Mei Ho House marks the history of early public housing in Hong Kong. The historical significance will mainly be preserved by limiting major alterations to the building structure. Furthermore, five units will be retained in their original state as sample rooms for the museum (Commissioner for Heritage's Office 2010b). Visitors will be given the opportunity to witness a piece of Hong Kong's important and unique historical development. Figures 9.2 and 9.3 illustrate the new Mei Ho House from an artist's perspective (Commissioner for Heritage's Office 2010e; 2010f).

Figure 9.2 Artist's impression of the future Mei Ho House's front view (Commissioner for Heritage's Office 2010e) (with permission from Hong Kong Youth Hostels Association)

Figure 9.3 Artist's impression of the future Mei Ho House's original room (Commissioner for Heritage's Office 2010f) (with permission from Hong Kong Youth Hostels Association)

Social benefits

The key social benefits that will be achieved include increased job opportunities, uplifting tourism, providing cheaper accommodation for visitors, providing new and alternative entertainment facilities, and also educating the general public on the history and culture of Hong Kong. It is anticipated that 290 jobs will be created during the renovation period and 42 full-time and 63 part-time jobs for local residents upon the project being commissioned. It is estimated that the new hostel and museum will attract 104,000 visitors per year for the first three years, with the accommodating ratio being 50 per cent, 65 per cent and 70 per cent in the first, second and third years respectively (Commissioner for Heritage's Office 2010b; Ming Pao 2010).

Public opinion

A spokesperson from the Development Bureau advocated that the price of staying at the newly developed hostel would be similar to that of two- to three-star hotels locally. The hostel will provide cheap accommodation for visitors. It is suggested that previous residents of Mei Ho House will be invited to introduce the buildings to visitors, so that they can better understand the history and culture behind Hong Kong's early public housing.

Initially the service provider will be awarded a '3+3' year contract, where the first three years will be fixed term and the latter three-year contract will be subject to the service provider's performance. The profit gained from the building will be used solely for its daily operation. It is anticipated that the project will break even after two years of operation. The related departments have been praised for doing a good job in making the project transparent as well as informing the general public of the details and progress (Hong Kong Commercial Newspaper 2010).

With the redevelopment underway in the Sham Shui Po district, Mei Ho House will be adjacent to a number of newly built private residential flats (Oriental Daily 2010a). Other developments confirmed include a historic building on Prince Road West to be developed into a hotel and an innovation and arts centre in Shek Kip Mei (Ming Pao 2010). These projects will complement each other to uplift historical preservation and cultural development.

Previously, there were concerns from legislative councillors as to whether the operation of these projects will need to be subsidised by local government in the future. But the Development Bureau is convinced that the projects would be able to run by themselves (Tai Kung Pao 2010a). Reports indicate that the government subsidy required for the project is necessary largely because of the poor structural state of the buildings. They were constructed quickly in an attempt to provide public housing for those who were homeless from the Shek Kip Mei fire, hence the quality was neglected. In addition, the buildings are over fifty years old, hence restoration has been foreseen. It is predicted that at least half of the money will be spent on reinforcing the foundations and strengthening the structure (Oriental Daily 2010b). It is estimated that the hostel alone will bring in HK$1.2 million (approximately US$0.2 million on 11 December 2012, Yahoo! Finance 2012) income in the first year (Tai Kung Pao 2010b).

The North Kowloon Magistracy case study

Background of the North Kowloon Magistracy

The North Kowloon Magistracy is the first building under this scheme which has been fully revitalised. The building was one of the first selected historic buildings amongst the first batch of the revitalisation scheme. The building is located at 292 Tai Po Street, Shek Kip Mei, which is in a central location of the Kowloon Peninsula of Hong Kong. The building served the Kowloon district community as a magistracy during 1960-2005 (Leisure and Cultural Services Department 2012). The Magistracy was finally closed down as part of the local government's cost saving policy by reducing the number of magistracies in Hong Kong (Antiquities and Monuments Office 2012). The building contained four magistrates' courts, one juvenile court and government offices on the upper levels. The Magistracy handled minor offences such as prostitution, littering, traffic offences, and so on. The maximum imprisonment given at this magistracy was two years and the maximum penalty was HK$100,000 (approximately US$13,000 on 11 December 2012, Yahoo! Finance 2012). During its service there were between forty and eighty defendants attending court on a daily basis (Judiciary of the Hong Kong Special Administrative Region 2012).

Background of Savannah College of Art and Design

The Savannah College of Art and Design (SCAD) was selected as the service provider for the North Kowloon Magistracy revitalisation project. It is an American institution providing education in topics which were not readily available in the area of the arts back in the late 1970s (Savannah College of Art and Design 2012a). The first campus in Savannah comprises seventy facilities in an area of over 2 million square feet, serving 8,000 students (Savannah College of Art and Design 2012b). The Savannah city is a renowned national historic landmark and SCAD has taken advantage of this by revitalising historic buildings within the city for its college facilities. Their success in adaptive reuse and urban revitalisation has been recognised by the Historic Savannah Foundation, the Georgia Trust for Historic Preservation, the Art Deco Societies of America, the National Trust for Historic Preservation, the International Downtown Association, the Victorian Society in America and the American Institute of Architects (Savannah College of Art and Design 2012c).

SCAD currently offers a range of programmes including the Master of Architecture, Master of Arts, Master of Arts in Teaching, Master of Fine Arts, Master of Urban Design, Bachelor of Arts, Bachelor of Fine Arts, graduate certificates and undergraduate certificates (Savannah College of Art and Design 2012a). In addition, SCAD has expanded to four campuses including Savannah and Atlanta in the United States, Lacoste in France and Hong Kong in China.

The revitalisation process

For each heritage building selected under this revitalisation scheme, a detailed resource kit was prepared by the Hong Kong government. The resource kit provides information regarding the project's background, administrative procedures and guidelines, and also technical issues. These resource kits which are readily available online allow potential service providers to consider their interest in the projects. An important component of the resource kit is that it states the features of the project that need to be preserved and in some instances the recommended treatment method is also indicated. Table 9.1 shows a summary of these for the North Kowloon Magistracy (Commissioner for Heritage's Office 2012).

This comprehensive list mainly includes those features to be preserved, but other features to be removed are also included. In general, the Hong Kong government requested few changes in order not to destroy the heritage value of the building and its surrounding area. The preserved features included those that have given the building its identity. Some of the main features preserved include: one of the courtrooms and its leading staircase; a detention cell with its stone bench and railings; the main entrance staircase and railings; the stone flooring of the foyer; the marble cladding of the walls and columns; the central staircase and railings; the window frames, balconies, handrails and guardrails; the doors and frames; the central lightwell; the canopy and flagpole; the signages on the building front; and the garages. Most of these features were just cleaned to avoid damage.

Table 9.1 A list of building features and their recommended treatment as requested by the Hong Kong government

Features	Recommended treatment
External	
Facade of building	Clean with water and undamaging tools, no corrosive cleaning chemicals, do not install protruding structures
Main entrance staircase and railings	Repair and clean with appropriate cleaning agent, modification works to railings to meet with safety requirements allowed but must be reversible
Name tablets at main entrance staircase and building front	Keep original or new letterings, change should be reversible
Road signages on exterior walls	Remove, clean and repair walls
Main entrance door and frame	Repair and clean with appropriate cleaning agent
Side entrance door and frame	Preserve
Metal-framed windows and balconies	Repair and clean with appropriate cleaning agent, modification works to railings to meet with safety requirements allowed but must be reversible

Continued overleaf

Features	Recommended treatment
Canopy and flagpole	Repair and clean with appropriate cleaning agent, do not install structures covering flagpole and base
Garages	Preserve
Two-storey temporary structure and fire services room	Remove
Two trees at south-east car park entrance	Preserve
Internal	
Painted walls	Preserve
Stone flooring of foyer	Preserve
Central staircase, railings and ceiling lights	Upgrading work of railings to meet current standards
Window handrails and guardrails	Preserve
Marble cladding of walls and columns	Preserve
Door openings and frames of staff canteen and shroff office	Replace with salvaged panelled doors
Wooden panelled doors and frames of offices, toilets, police general registry, summons office, changing room, courts 1–4, staff canteen staircase, meters and switches room	Reuse panelled doors
Doors and frames of police duty room, lavatory and reception room	Preserve
Concrete benches and iron bars of one detention cell	Preserve
One courtroom and the staircase leading to it	No additional floors within courtroom
Central light well, roof light, exhaust fan housings, windows, guardrails of internal corridor	Preserve
Safe no. GSD1297 and keys, old furniture, equipment, plaques.	Return to government
Notice boards, built-in furniture/furniture, partitions, etc.	Remove and renovate wall surfaces
Wooden floorboards	Sand and wax or remove and replace
Internal walls and partitions	Non-load-bearing walls can be removed
Ceilings	Remove false ceilings to check and repair roof leakage, reinstating false ceilings allowed
Redundant building services	Remove and rewire
Interior signages	Remove and renovate walls
Toilets and kitchen	Remove and renovate walls and floors

Source: Commissioner for Heritage's Office 2012) (with permission from the Commissioner for Heritage's Office, Development Bureau).

There were other features of the building which were believed to be unnecessary and were requested to be removed. These included the toilets; kitchens; signages; furniture; abundant fire services; and temporary structures outside the building.

Site visits

A total of 10 participants were invited to join a guided tour of SCAD Hong Kong (SCAD HK) (Figure 9.4). The participants included academic, administrative and technical staff as well as postgraduate and undergraduate students from a local university. Due to the large number of participants, the group was divided into two groups to visit the site on separate occasions between May and June 2011. The guided tours lasted 45 minutes each and were led by the admissions staff of SCAD HK. As part of the revitalisation scheme's requirement, the service provider is responsible for engaging and involving the general public; consequently SCAD HK offers guided tours which are open to all visitors given that advanced booking is made. During the tours the participants were briefed about the historical and preserved features of the building as well as those new additions which had been included to provide for the facilities necessary for the building's new function. Some of the highlights of the tour included the old courtrooms which had been transformed into a lecture hall (Figure 9.5), student work station and room for photography shoots; the juvenile courtroom which had been converted to a library; the car parking areas which have retained the same use; the preserved cell (Figure 9.6) and the cells which had been converted to small study rooms, offices and storage; the old office areas which have been converted to classrooms; the old kitchen and canteen which had been converted to an art gallery (Figure 9.7); and the preserved foyer, open areas and staircases (Figure 9.8). Other information, such as that related to the college and programmes was also provided to the participants during the tour. The participants were also given the opportunity to ask additional questions regarding any other aspects.

Focus group meetings

Focus group meetings are a convenient, effective and fast way to collect a vast amount of information from a reasonable number of participants compared to the traditional one-to-one interview technique. Focus groups have been found to provide a highly effective and efficient way of investigating (Haslam 2003). Vaughn *et al.* (1996) believe that focus groups should possess two core elements: (1) a trained moderator who sets the stage with prepared questions or an interview guide; and (2) the goal of eliciting participants' feelings, attitudes and perceptions about a selected topic. In a focus group meeting, it is best that there are at least six participants in a group (Morgan 1997). The reason is that with fewer than six it may be difficult to sustain a discussion. The groups are given an introduction: this part of the focus group meeting is vital to its success. The typical introduction generally includes a welcome, an overview of the topic, the guidelines for the discussion, and the opening question (Krueger and King 1998). The interaction between group members is known to be an effective way of obtaining adequate information.

Figure 9.4 Front view of SCAD HK (with permission from Dr Bingqing Zhai)

Figure 9.5 Old courtroom transformed into a lecture hall (with permission from Dr Bingqing Zhai)

Figure 9.6   Preserved cell (with permission from Dr Bingqing Zhai)

Figure 9.7   Old kitchen and canteen converted to an art gallery (with permission from Dr Bingqing Zhai)

Figure 9.8   Open areas and staircases (with permission from Dr Bingqing Zhai)

This technique was used to gather information regarding the effectiveness of the scheme via the SCAD HK project. Amongst the participants who joined the site visits, sixty-six attended the focus group discussions. The participants were arranged into groups of approximately ten participants (Figure 9.9). Each group of participants was led by an academic facilitator. The facilitator introduced the respondents to the format of the focus group, the items which would be discussed, and also took the responsibility for initiating a discussion amongst the respondents regarding some predefined questions. The questions were based on three major areas including the effectiveness of the revitalisation scheme, the impacts of the scheme, and also the extent of general public participation. The discussions lasted approximately forty minutes for each group. The findings from the discussions are presented in the following section of this chapter. The seven questions according to the three areas are listed below.

- Effectiveness of the revitalisation scheme:
  1   Has the revitalisation scheme been effective in conserving and adaptively reusing the former North Kowloon Magistracy building into the Savannah College of Art and Design?
  2   Has the revitalisation scheme been able to preserve the heritage value of the former North Kowloon Magistracy building?
  3   Was a balance between heritage conservation and adaptive reuse achieved via this scheme?

Figure 9.9 Focus group discussion (with permission from Dr Bingqing Zhai)

    4   The local Development Bureau had originally wished that SCAD HK would become a landmark, has this been achieved?

- Impacts of the revitalisation scheme:

    5   What other purposes could this building have been used for? For example, a community education base, a site to protect the local collective memory etc.

    6   What impacts or significance does this project have on the local community or Hong Kong society? Please discuss according to the social, cultural and heritage conservation perspectives.

- Public participation:

    7   Have the local community been active in participating in this project effectively? Considering the diversified needs and requirements of the local community, how can they participate in this project?

Focus group findings

Effectiveness of the revitalisation scheme

The participants were asked four questions regarding the effectiveness of the revitalisation scheme based on their observation from the SCAD HK project. The first question they were asked was: 'Has the revitalisation scheme been effective in conserving and adaptively reusing the former North Kowloon Magistracy building into the Savannah College of Art and Design?'

Amongst the six groups of respondents, the majority, five groups, responded positively agreeing that the project building was well conserved and reused. Group A commented that the building structure was revitalised very successfully.

Nevertheless, they also mentioned some drawbacks including the small library and the high tuition fees. Similarly, Group B also mentioned that the building was well utilised and that little damage had been done to the building. They also commented that the original building was well suited for the new purpose and that the tour was interesting to visitors. Group C was the only group with comments solely negative towards the project. The participants of this group suggested that there was limited interaction between SCAD HK and the local community and that visitation would be impossible without going through a complex registration process. Group D agreed with Groups A and B in that the project was successful in conserving heritage and also that the original building was suited for the new purpose. However, Group D also agreed with Group C that interaction between SCAD HK and the local community was lacking. The respondents felt that SCAD's understanding of the local community was very different from reality. Group E also felt that the project was well revitalised in aspects such as the stone flooring. There are certain features that they felt could be better preserved such as the sanitary facilities which had been modernised. In addition, they made some further comments to improve other revitalisation projects. For example, building requirements for revitalisation projects could be made more flexible as some old features would be difficult to comply with new standards. Another example is that more information regarding the preserved and historic elements could be provided and advertised to the general public, as well as the activities offered by the service provider. Similar to Group C, they also mentioned that a key limitation is that walk-in visitation was not permitted without arrangements being made beforehand. Furthermore, revitalised elements should be reversible where possible. Group F also agreed that the project was successful in terms of conservation. They appreciated the fact that the original elements were well preserved, in that outsiders would not notice it was functioning as an art college. They also supported the reuse of the building so that it could serve a new function. But unlike other groups the respondents felt that the interaction between SCAD HK and the local community was also successful.

Regarding the effectiveness of the scheme, the participants were asked to discuss a second question: 'Has the revitalisation scheme been able to preserve the heritage value of the former North Kowloon Magistracy building?'

In general, the participants felt that the historic features of the project were well preserved. Group A felt that the scheme has been very successful in preserving the original features. Their only concern was that the project may not be financially sustainable in the long run due to the low student enrolment. Group B agreed that the historical significance had been maintained due to the minor alterations. Similarly, Group C agreed that the building had been well preserved. Group D suggested that revitalisation should be considered across different industries and that integration should be achieved. Group E discussed that there are two elements which can be preserved: the hardware and the software. From the hardware perspective they agreed that the physical structures have been well preserved. But from the software perspective the project has not been conserved for its original use. The participants added that the project was supposed to merge with the local community but they felt that interaction between SCAD HK and the local community was lacking, and that the project was very different from the surrounding area.

Group F agreed that the heritage conservation of the project was carried out very well, due to the sizeable financial support from SCAD. Small features were also preserved and the features in general matched the new function. They felt that the building was well suited for the new purpose as an art college.

For the third out of four questions on effectiveness, the participants were asked: 'Was a balance between heritage conservation and adaptive reuse achieved via this scheme?'

The participants did not agree or disagree strongly whether a balance between conservation and reuse had been achieved. Instead they discussed the different perspectives which surround this topic. Group A did not feel that the new use was inappropriate but felt that it could be better utilised as a law school so that more of the original features could be reused. Nevertheless they felt that the new building service facilities were cleverly incorporated, bringing in modern features to an old building. This group also discussed that there were two modes for financially supporting reuse of buildings. The first mode is where the majority of funding comes from the government, such as the Forbidden City in Beijing, China. The second mode is for non-governmental organisations to support the project by running a business out of it like SCAD HK so that they are self-financing. The participants did not come to a conclusion as to which mode would be better but just highlighted that there are two options for how reusing buildings could be financed. Group B suggested that a balance between conservation and reuse is difficult to achieve. In addition, they suggested that the project would be difficult to maintain in the long run because of the minimal works conducted. Maintenance would be much easier for conservation projects where more new finishes are incorporated, such as the 1881 shopping mall in Tsim Sha Tsui of Hong Kong. For the SCAD HK project, Group C felt that the local government was unclear about the scheme. They questioned whether the project should be considered a conservation project or a redevelopment one, as they felt that SCAD HK had few local interactions and much of its reputation was achieved overseas. Group D shared the same view as Group A, that the project would be better utilised as a law school to synergise the original purpose with the new one. Group E discussed that each party would have a different view on whether a balance between conservation and reuse had been achieved. From SCAD's point of view it has, but from the local residents' angle it may not have. Group F was the only group to think conservation and reuse had been well balanced. They discussed that the building was well preserved and that the project was self-financeable.

The participants were asked a final question regarding the effectiveness: 'The local Development Bureau had originally wished that SCAD HK would become a landmark, has this been achieved?'

The majority of views suggested that SCAD HK has not become a landmark locally. Group A suggested that there were several reasons why SCAD HK has not become a landmark. First, its location itself dooms it from becoming a landmark, as it is located on the outskirts of an old town area. Second, there has been little government publicity regarding this project. Also, the art college has a high-end image which may not match with the local grass-roots community. The participants suggested that a local service provider such as the Chinese Artists

Association may have been more appropriate for the area. Group B also agreed that as SCAD HK was located on the outskirts of the town it was difficult for it to become a landmark within the area. The participants felt that people would consider the better known computer shopping malls in the area. Nevertheless, the participants still thought highly of the project and believed that it would be recognised as an example of a successful revitalisation project on an international level. Group C suggested that the project has little connection with the local community so it cannot be considered a landmark. Group D shared similar feelings to Groups A and B that the location of the project is too remote to become a landmark. Group E did not decide whether SCAD HK had become a landmark, but instead suggested that a good example has occurred in Newcastle in Australia where an industrial site had been transformed into a recreational venue and at the same time maintaining many of its original features which have attracted many visitors. The participants of Group F shared mixed views, where some felt that the project had become a landmark but others disagreed. For those who felt that it wasn't a landmark, they felt that both the original and current functions did not draw many visitors, also they felt that the current usage did not merge well with the local area.

Impacts of the revitalisation scheme

The participants were asked two questions regarding the impacts of the revitalisation scheme. The first question they were asked was: 'What other purposes could this building have been used for? For example, a community education base, a site to protect the local collective memory, etc.'

Some of the participants supported the building being used as an art college whereas some did not, but it was agreed that the building could have been better utilised for purposes related to its previous functions. Group A did not feel that this building could serve as a site of collective memory for the local residents as its former function was not a place that they would have visited. They added that there is a Chinese saying that 'you would never go to court when you are alive just like you would never go to hell when you die'. With this mindset the participants felt that the local community would not have visited the site voluntarily hence there would be no collective memory worthy of preservation. Group B supported the fact that the site had been revitalised as an art college but they felt that improvements could be made to enhance community participation. For example, they felt that the introductory tours could be made more flexible so that visitors would be welcome throughout the day rather than having to make prior arrangements with SCAD; this would allow more flexibility for interested visitors. The participants also valued the introductory videos which were shown but felt that the information provided in them was not detailed enough in terms of the revitalisation process, instead much information was based on SCAD's activities and background. An example of a good introductory video is the one produced for the Dr Sun Yat-sen museum in Central of Hong Kong. Furthermore, the art gallery was unattractive as the majority of Hong Kong locals are not particularly interested in arts and culture, plus the gallery is small in scale and holds few exhibits. Group C also supported the new usage of this building and felt that it has set a good example for other revitalisation projects in Hong Kong. But they felt that the original and new usages of historic buildings should not differ drastically. For example, they supported the idea of Murray House in Hong Kong, which was an old officers' barracks to be transformed to the Hong Kong Maritime Museum, but they did not support the idea of allowing restaurants on the upper floors of the building. Group D felt that the current usage as an art college could not benefit the local residents; instead they suggested that it could have been reused as a museum, library, or a building with a similar public function. Similarly, Group E agreed with Group D that the current art college drew few local residents because of the high tuition fees and high-level image. Instead they felt that a Chinese opera house would arouse more interest amongst the ageing community which it is within. Group F's comments were similar to that of Group C's in that they believed the previous and current usages of the building should not differ too much. Their suggestions included a museum for Hong Kong's legal system or a mock courtroom for Hong Kong law schools. They added that the new function of a historic building should be considered on a case by case basis.

Regarding the impacts of this project, the participants were also asked: 'What impacts or significance does this project have on the local community or Hong Kong society? Please discuss according to the social, cultural and heritage conservation perspectives.'

In general, the participants felt that there were few, if any, impacts or significance generated from this project. Group A felt this was due to the minimal marketing conducted. The participants suggested that photos of the original building could be displayed inside for visitors to compare with the current state. From the economic impact perspective, the participants felt there was limited impact derived from the staff and students of SCAD. From the cultural and heritage conservation perspective, the impact affected the staff and students more than the local community or Hong Kong society as they get to know the culture of Sham Shui Po better. Group B once again echoed that the local community would not benefit from SCAD HK as the tuition fees would be too high for them to afford. Even for those Hong Kong residents who could afford the tuition, they may consider studying elsewhere or even overseas. Again, the location was a primary factor causing minimal interaction between the local residents and SCAD UK. Regarding the economic benefits, they also felt that these were minimal as academics were drawn from the United States rather than locally, hence few employment opportunities have been created. On the contrary, SCAD has benefited in providing their students with another campus. Although so, they feel that this project has bought a positive image to Hong Kong in preserving historical buildings and reusing them for something other than shopping centres or restaurants which has been the more common practice adopted in Hong Kong. These previous projects were more commercial rather than focusing on the preservation of the buildings. Therefore, they felt that the reuse as an art college was innovative and should be supported. Having said that, the participants still felt that they and even the local residents would have preferred a local service provider with which they could better connect. Group C simply reflected that there was limited impact to the local community as SCAD HK could not serve its needs. Group D believed that the project's social and economic impacts were limited because of the remote location of the building and the limited employment opportunities derived from the construction and operation. Regarding the heritage aspect, the participants felt that the current introductory tour about the former usage was informative; they suggested that it would serve as a good base to educate the local community and students about heritage conservation. Group E agreed with the majority that there was little social impact created from the project; they believed that the local government should have encouraged this by requesting SCAD HK to provide either free or cheaper short courses for the local community. Another suggestion is to encourage co-organised courses between other local educational institutions at discounted prices. These requests could have been incorporated as a condition within their agreement. In addition, this group was the only one that believed economic benefits could be created from the business opportunities for the local restaurants, shops and hotels. Group F did not have any comments regarding this question.

Public participation

The participants were asked their views regarding public participation of the revitalisation scheme via the following questions: 'Have the local community been active in participating in this project effectively? Considering the diversified needs and requirements of the local community, how can they participate in this project?'

Group A felt that the local community are aware of this project or the other revitalisation projects; regular visitations should be made available for the locals. Some of the participants within this group felt that public participation would be better achieved if the project was used as a shopping mall rather than an art college. Similar to Group A, some of the respondents of Group B supported the idea of similar projects being reused as shopping malls to draw public participation. They felt that projects which have been reused as shopping malls attracted more visitors, whereas SCAD HK benefited their students only. Nevertheless, they felt that the location of SCAD HK would not be convenient as a shopping mall as it was too remote. Their conclusion was that only historical buildings within the city centre should be reused as shopping malls. Therefore, they felt that it was wasteful for the government to spend money on buildings not in the city centre and where few visitors would be attracted. Group C suggested that more promotion materials should be produced to attract visitors such as a small booklet describing the project. Group D agreed with the other groups that public participation was lacking. The participants suggested some measures to increase the general public's interest including inviting the local community to see the exhibitions, arranging guided tours for the local community, and SCAD HK staff and students could combine local craftsmen's art works with their works. Group E added a further suggestion for improving public participation by easing the visitation process for the local community. Group F shared an interesting view that introducing the previous occupants of the building would ease public participation, but in this case it may not be feasible.

Chapter summary

This chapter has looked into the effectiveness of revitalising historic buildings through a partnership scheme introduced in Hong Kong. The Mei Ho House was used as the first case study to investigate the effectiveness of this scheme. The results indicated that the project to date has been well received by the general public. No significant dissatisfaction, problems or complaints have been recorded. Therefore, the analyses confirm that this scheme is a feasible method for revitalising historic buildings. The success of this method might bring opportunities for other historic buildings needing to be preserved. This scheme has been particularly successful in Hong Kong as the local government is comfortable and happy to fund these projects at the initial stage. With the expertise and skills from the private sector, these buildings can be better utilised and preserved. As a result, the scheme will help to maintain history and culture, draw visitors, educate the general public, uplift the image and attraction of Hong Kong, introduce job opportunities which will in turn improve the local economy, and also provide alternative and more entertainment facilities and services to both visitors and the locals. Obviously, this method can be applied to other jurisdictions which have a similar situation to Hong Kong, where financial issues are not the government's primary concern, but where the aim is to better preserve and make use of historical buildings.

The second case study analysis was the Savannah College of Art and Design. This project is the first completed project under the Revitalizing Historic Buildings Through Partnership Scheme, introduced by the Hong Kong government. Participants were invited to take part in a focus group discussion, in order to provide their views regarding the effectiveness, impacts and extent of public participation of the scheme via this project.

Regarding effectiveness, in general the respondents reflected that the project building was revitalised successfully and that the new purpose was suited. Nevertheless, they also reflected that some of the drawbacks included the lack of interaction between SCAD HK and the local community, and also the need for prior arrangements to visit. During the focus group discussions, some suggestions were also provided by the respondents to improve future similar projects. These included allowing more flexibility for building requirements of revitalisation projects as old features can be difficult to comply with new standards. Projects can be better advertised so that the general public can get to know the historic building and its available services. Revitalised elements should be reversible where possible. In general, the participants agreed that the heritage of the building was well preserved. Regarding the balance between conservation and reuse, the participants expressed mixed views. The majority of participants did not feel that the project had become a landmark for the Hong Kong people mainly due to it being a high-end organisation in an old grass-roots community. Also, the location is not convenient for visitors. But the participants did suggest that the project would be renowned on an international level for being a successfully revitalised building.

The participants also discussed the impacts as a result of the revitalisation scheme in the case of SCAD UK. The discussion showed that the impacts created were not huge but the revitalisation scheme is innovative and has helped uplift

Hong Kong's image in heritage preservation. The building was not believed to hold many collective memories as it was not a place which the general public would visit often during its original usage. The participants recommended that the new and original usages of historic buildings should be related so that the revitalisation becomes more meaningful and so that the buildings could be better utilised. It was felt that minimal impacts have been created due to the lack of marketing, the remote location and the high tuition fees.

Regarding the aspect of public participation, it was felt that this was limited due to the reasons discussed. Recommendations for improving public participation included to arrange visitations especially for the local community to see the exhibitions and participate in the guided tours, to produce promotion materials, and for SCAD HK to work with local craftsmen. Nevertheless, the revitalisation scheme has still been regarded as an effective approach in reusing and preserving historical buildings. Consequently, the image of Hong Kong in the area of heritage development has been uplifted on an international level.

10 Learning from less successful cases

Introduction

This chapter presents four case studies analysed by previous researchers. These public–private partnership (PPP) projects were all heavily criticised in some way for not meeting expectations. They include the Southbank Education Training Precinct in Brisbane, Australia (Chan *et al.* 2008a), the Sydney Cross City Tunnel in Sydney, Australia (Chan *et al.* 2008b), the West Kowloon Cultural District and the Western Harbour Crossing, both in Hong Kong (Chan *et al.* 2007b).

The Southbank Education Training Precinct

Background

The first proper PPP project conducted by the Queensland government was the South Bank Education and Training Precinct (SETP) in Brisbane, Australia. In September 2002 the development of the SETP was announced. The goal was to develop a multi-sectoral campus that built on developing greater links between schools, technical and further education (TAFE), universities, community groups and industry.

The private sector was invited for expression of interest from February 2003 to 11 April 2003. Three consortia were shortlisted by August 2003. In the end only two bids were received, and Axiom Education Queensland Proprietary Ltd was selected as the more favourable consortium in December 2004. The parties within the consortium consisted of ABN Amro, John Holland Proprietary Ltd and Spotless Services Australia Ltd. After further rounds of reviews, especially on the value for money aspect, the consortium was announced successful on 19 April 2005, to plan, design, construct, finance and maintain the AU$550 million facilities. The construction of this four-hectare site was completed in 2008. The concession period will be for thirty-four years (Queensland Government 2008b).

The SETP was initiated by the then Department of Employment and Training in consultation and collaboration with the Department of Education and central agencies, including the Premier and Treasury. As the SETP represented the Queensland government's first project to be commenced under PPP guidelines, there were no state-based or local examples to go by. Therefore, in order to aid the concessionaire, they were given access to policy material from the Department of State Development and Innovation and also the opportunity to interview inter-state government agencies which had direct experience on the development and procurement of PPP projects. It was in this way that the concessionaire of SETP learned to conduct PPP projects. In addition, the selected consortium already had previous experience in PPP projects elsewhere.

The obstacles

Being the first PPP project in Queensland, SETP faced a number of problems early on. First, the government's design of the project was very conceptual. Therefore the actual validity of the public sector comparator was highly questionable and hence so was the debate on value for money. However, the public sector comparator was still conducted just as a governmental process to justify the project being a PPP. Another worry for the private consortium was the Queensland government's commitment to proceeding with the project. It is known that some organisations and companies in the private sector were reluctant to participate in this project as the state government has had a previous track record of pulling out of a proposed PPP project. With the large project sum and the complexity of PPP projects, the risks that they bear are immensely high. Therefore, the private sector needs to be sure that they have the government's full support throughout the project to overcome potential obstacles.

In addition, many experts in the field believed that the Queensland government should not have chosen such a complex project for its first PPP project. Social projects are often considered more difficult to handle compared to economic ones where income is obvious. In addition, partnership projects between the public and private sectors have always traditionally been economic ones hence the experience and knowledge in this field is much more advanced. This was another reason that drew away potential bidders, because they lacked confidence that the project would succeed. In the end there were only two bidders for this project. The lack of potential bidders is another issue thought to limit the competition in a PPP project. The lack of competition can be linked to the lack of quality, innovation and price delivered. In order to boost up the number of potential bidders in future, not only does the first project need to be a success, but the Queensland government also proposed a small amount of compensation to the losing bidder. In the SETP case the losing bidder was compensated AU$3–4 million (approximately US$3.1–4.2 million on 11 December 2012, Yahoo! Finance 2012). Although this is not the full amount that they would have spent during the tendering stage it was comforting to get a partial return of the investment.

For the SETP project many of the protocols were already set out in the project deed issued by the Queensland government. In terms of management practice there was little difference from conventional methods. But a more rigorous risk management practice and a better management practice overall were incorporated. Being a PPP project the other issues that the consortium needed to deal with were the management of publicity and media issues. In addition, the initial phase of this project took a long time, much longer than it should have. The government also noticed this and has stated that for future projects they would definitely need to streamline the timeframe.

The Sydney Cross City Tunnel

Background

The Cross City Tunnel (CCT) in Sydney, Australia is a good example of how improper allocation of risks could affect the success of a PPP project (Figure 10.1 is a photograph of the CCT). It is not incorrect for risks to be passed on to the private sector, especially when they are capable of handling them. But maybe there should be a 'partnership' in place when the private sector is unable to manage all the risks alone. Some critiques considered this project to be an unsuccessful PPP as the local state government has had to cope with handling much public criticism for its inaccurate traffic forecasts, leading to the investor making a financial loss.

The primary objectives of the CCT project were to reduce 'through' traffic in central Sydney, and as a result easing traffic congestion and improving environmental amenity in the central business district and on streets approaching the central business district, and to improve the east-to-west traffic flows (Roads Traffic Authority 2007).

Figure 10.1 Photograph of the Cross City Tunnel (Judd 2008) (with permission from Trafford Judd)

The CCT is a 2.1-kilometre twin two-lane motorway that runs east and west underneath the busy central business district of Sydney. It opted for a Design Build Operate (DBO) arrangement under a thirty-year concession agreement. The project was part of a network of new transportation infrastructure plan of the Roads and Traffic Authority of the New South Wales government. Its large project sum of AU$680 million meant that PPP was an attractive option to the New South Wales government.

The initial concept of the tunnel was mooted in 1998 (Cross City Tunnel Proprietary Ltd 2007). After a series of complex consultations, exhibitions, modifications and approvals the private sector was finally asked for an expression of interest on 15 September 2000 (Roads Traffic Authority 2003). In response, a total of eight consortia expressed interest by 23 October 2000. Three consortia were shortlisted and asked on 8 June 2001 to submit detailed proposals for the project. All three consortia submitted their proposals by the closing date of 24 October 2001. It was announced on 27 February 2002 that Cross City Motorway Proprietary Ltd was selected as the winning consortium.

The project commenced construction on 28 January 2003. It took only thirty-one months to construct and was delivered ahead of schedule (typical for PPP projects). The tunnel was officially opened for service to the public on 28 August 2005. Unsurprisingly, the project attracted private sector interest from within Australia and abroad. The selected consortium included strong financiers: Cheung Kong Infrastructure of China, Bilfinger Berger of Germany and RREEF Infrastructure of Australia. They would bring in equity and recover the cost of design, construction, operation and maintenance via the tolls collected. Therefore the project company Cross City Motorway Proprietary Ltd was allocated all the demand risk for the project. Innovation was introduced by the contractor. The tunnel was the first motorway in Sydney to have full electronic tolling. There were high expectations by all the parties; the traffic forecast for the project was predicted to be 90,000 vehicles per day.

A number of benefits were sourced from materials published and released from the project company Cross City Motorway Proprietary Ltd (Cross City Tunnel 2007) and the government agency client the Roads and Traffic Authority of New South Wales (Roads Traffic Authority 2007). These parties claimed that as a result of the Cross City Tunnel project the following benefits would be experienced:

- Thirty-four traffic signals avoided (sixteen sets westbound and eighteen sets eastbound);
- Major reduction of traffic across the central business district;
- Improved quality of life for pedestrians and cyclists in the central business district;
- Higher reliability of bus services in the central business district;
- Cut trips across the city to approximately two minutes, from up to twenty minutes by avoiding traffic lights;
- Improved access and movement within the city for taxis, delivery vehicles, cyclists and pedestrians;
- Make city streets safer and more pleasant for pedestrians, residents and

business people by removing intrusive through traffic and providing more
footpath space in some streets;

- Reduced traffic noise levels;
- Better air quality by taking cars off surface streets.

Despite the benefits of PPP which have been highly publicised, some may consider
that there are also many 'failures' from the project. The next sub-section takes a
closer look into these 'failures'.

Underlying causes leading to failure

CCT has been perceived as an unsuccessful case by the general public and as a
result the state government's image has suffered (Jean 2006). To illustrate some
of the negative portrayals of the project, some headlines related to the project
were sought from Infrastructure Implementation Group (2005). They included:
Tunnel cuts William Street to one lane to trap drivers; Cheap tunnel buyback
mooted; $105m toll outrage; Cross City grovel – Three weeks toll free but roads
still clogged; Taken for a ride – Tunnel at the crossroads; Changes to contract led
to high tolls; and Drivers to feel squeeze. Amongst these seven headlines, three are
related to the toll. This can show that the toll is probably one of the key factors
affecting the satisfaction level of the general public towards CCT, and also one of
the issues that is highly sensitive amongst them.

PPP has been given a bad name and investors have been driven away from New
South Wales, at least temporarily (AAP General News Wire 2006a). CCT encountered severe difficulties in reaching the predicted traffic volume. Motorists have
expressed their unhappiness about the high toll levels (AAP General News Wire
2006b) and the government closing off the surface roads to divert the traffic into
CCT (AAP General News Wire 2006c). These sufferings have been the result of
inaccurate traffic forecasts and a flawed concession agreement. Currently, CCT
has entered into receivership and the concessionaire has written off their equity
(Project Finance 2007).

In this project it has been unfortunate that the public client and the private
consortium have openly argued in public. Newspapers have reported them criti-
cising each other for their faults (Field 2006a). The Premier publicly spoke out
expressing his frustration that motorists were able to use the toll road without
paying. He criticised the operators for not enforcing the charge and how it was
unfair for the motorists who did pay (AAP General News Wire 2006d; Field
2006b). On the other hand the consortium also criticised the Premier for failing
to demonstrate leadership (AAP General News Wire 2006e). It can be seen how
the media has portrayed a tense battle between the public and private sectors. This
is an image that nobody wants to create for any project whether it is delivered by
PPP or not. But being a PPP project creates an even higher sensitivity, as taxpayers
will query whether they are actually getting value for money from the government's
decision.

Following the unfortunate events experienced, the private consortium requested
the local state government to pay them a toll subsidy and compensation for the road

changes. Unfortunately the two parties were unable to come to a satisfactory agreement (AAP General News Wire 2006f). But in order for the CCT case not to be repeated, the local state government considered paying the consortium compensation for the forthcoming Lane Cove Tunnel, which is also in Sydney, if traffic forecasts are also predicted inaccurately (Cratchley and Jean 2006a; 2006b). This action from the government was positive as it showed that they were aware of problems in the CCT project, and that they should share the responsibilities by undertaking more of the risks rather than passing the pressure solely to the private consortium.

In 2005 the New South Wales government produced a report titled *Review of Future Provision of Motorways in NSW* (Infrastructure Implementation Group 2005). The report reviews recent road projects including the CCT in order to improve future similar projects. It is unfortunate to see that more barricades are set up to protect the government, as a result of which risks are further passed on to the private sector. For example, in the document they expressed their preference to bidders with the 'lowest' toll. This line of thinking is similar to selecting the lowest-cost bidder, which should not be the only way to select the consortium. Instead, value for money for the project overall should be their main concern. By focusing on the toll only, other important features adding to value may be neglected, such as innovative techniques and skills used in the project to make it more efficient and as a result creating value for money. Quality of the works may also suffer.

In the report it was also mentioned that in the neighbouring state of Victoria, all the main variables which would affect the commercial outcome of the project for all parties would be negotiated at the bidding stage. But in New South Wales the toll level or the possibility of government contribution would not be open to negotiation. Therefore it is questionable whether value for money for the taxpayers is achieved. The report has indicated that the New South Wales government is clearly aware of its faults, but whether it actually rectifies the situation is to be observed.

To consolidate the findings reported by the press discussed previously, the underlying causes leading to the 'failure' of the CCT project include:

- Inaccurate traffic forecast;
- High toll levels;
- Government closing off the surface roads to direct the traffic into CCT;
- Flawed concession agreement;
- The public client and the private consortium arguing openly in public;
- No toll subsidy and/or compensation from the government;
- The toll level or the possibility of government contribution was not open to negotiation.

West Kowloon Cultural District

The proposed West Kowloon Cultural District (WKCD) project is located West of Yau Ma Tei on the Kowloon Peninsula of Hong Kong. Figure 10.2 shows an illustration of the project (Foster and Partners 2013). The project had been long awaited due to repeated delays and controversies (South China Morning Post 2008). Back in April 2001, the Hong Kong government invited the private sector to submit conceptual plans for the forty-hectare waterfront site at the southern tip of the Western Kowloon reclamation into an integrated arts, cultural and entertainment district under the mode of PPP (Mok 2005). Since then, there has been continuous debate over which would be the best financing modality to be adopted. A number of uncertainties in this project doomed it to be highly criticised. These problems included:

- Initially the government was unclear of what they wanted and their objectives were unknown;
- The government did not notify or seek advice from relevant parties at the beginning, for example from art and culture groups;
- There was no timeframe or schedule to be met and no milestones to be achieved;
- The private sector was kept in the process for too long causing loss in terms of time and resources;
- The project may not be possible without heavy financial support.

Media reports have also highly publicised the dissatisfaction heard from local Legislative Councillors towards the government's indecisiveness and delay in delivering this project (Leong 2008). Due to the large number of uncertainties involved, the local government has needed to handle the project more cautiously. As a result a decision on the project design, timeframe and delivery method took a long time to be made. A number of public consultations were conducted, and a special task force was established to monitor the progress of this project.

Figure 10.2 The proposed West Kowloon Cultural District project (Foster and Partners 2013) (with permission from Foster and Partners)

The debate over the financing modalities finally came to a halt when the Legislative Council of the Hong Kong government approved HK$21.6 billion (approximately US$2.78 billion on 11 December 2012, Yahoo! Finance 2012) to fund this project (Wu 2008). The agreement was reached after thirty-two Legislative Councillors agreed to this arrangement against ten that opposed.

Western Harbour Crossing

The success of the Cross Harbour Tunnel (CHT) introduced around a dozen more Build Operate Transfer (BOT) projects in Hong Kong. Other examples of local BOT projects include the Chemical Waste Treatment Plant on Tsing Yi Island, the Tate's Cairn Tunnel and the Route 3 Country Park Section. But not all these projects were equally successful. A typical example is the Western Harbour Crossing (WHC) opened in 1993. Figure 10.3 shows a photograph of the WHC (Forum Sara 2008).

This project was the third underwater roadway tunnel to connect Hong Kong Island with the Kowloon Peninsula. This project was constructed as part of a giant infrastructure improvement project reaching HK$160 billion (approximately US$20.6 billion on 11 December 2012, Yahoo! Finance 2012) in scale, centred on Hong Kong's new airport (Nishimatsu 2006). Under the contract agreement of a thirty-year period, the consortium can adjust the toll depending on the performance of the revenue. If the revenue is underestimated, the toll can be increased to meet targets; on the other hand the toll can also be lowered if the toll exceeds the expected revenue. When the tunnel came into operation in April 1997 (Mak and Mo 2005), Hong Kong was experiencing an economic downturn which in turn reduced the traffic volume. Another problem was that the WHC was very expensive to build. It cost approximately HK$7,500 million (approximately US$965.75 million on 11 December 2012, Yahoo! Finance 2012), which was over twenty-three times more than that for the CHT (Li 2003). Therefore, in order to reach target revenue, the WHC increased the toll causing drivers to use the other two cross-harbour tunnels linking Hong Kong Island to Kowloon Peninsula (Kwan 2005).

WHC can therefore be described as a less successful project. The project investors have not met their target revenue, the general public has a negative perception of the project due to some adverse media reporting, and the local government has had to take criticism from the general public.

Analysis conducted by Tam (1999) showed that although the WHC was also completed earlier and effectively managed, the project performance was destroyed by the delay in toll rise applications. The applications were held up by the elected members of the Legislative Council of Hong Kong when the government introduced more democracy into the political system from 1990 onwards. The importance of a well-defined toll rise mechanism was crucial in this case. Political influence was also a major factor towards the end performance. As a result, an arbitration clause and a legal framework were established to resolve the dispute. Another possible factor affecting the difference in performance between the CHT and the WHC could be the lower government stake in the WHC. Government stake in projects could further ensure the commitment from all parties.

Figure 10.3 Photograph of the Western Harbour Crossing entrance and exit on Kowloon Peninsula (Wikipedia 2013)

Chapter summary

This chapter has presented four high-profile and criticised public works projects. Their experiences have resulted in some valuable lessons learnt for future PPP projects. It is important that governments and consortia can reflect on these lessons and make their decisions wisely in order to avoid the possible risks. A summary of lessons learnt from these projects are as follows:

- Governments should have clear objectives;
- The public sector comparator should be adopted properly to ensure value for money;
- The government's commitment is vital towards the project's success and to give the private sector confidence;
- Governments that are new to PPP should initially select straightforward projects;
- There should be competition between bidders to improve the overall quality;
- The setup process of PPP projects should not be lengthy;
- Positive media and the general public's support are vital for the success of PPP projects;
- Governments should ensure that any fees collected from the general public are reasonable and acceptable;

- Governments should share the responsibilities with the private sector in the event a risk occurs;
- There should be clear planning and scheduling;
- The government should respect the views of relevant parties before initiating projects;
- There should be demand for the project.

11 Public–private partnerships in the developing world

Introduction

This chapter looks at the use of public–private partnership (PPP) in the developing world via three case studies of the water industry in Ghana (Ameyaw and Chan 2012a; 2012b). The case studies illustrate that the water industry in Ghana reflects the typical challenges faced in developing countries for public utilities, where the government invests insufficiently in the projects. PPP has consequently provided a solution for providing these necessary public facilities and services where governments in developing countries are unwilling or unable to financially support. Unfortunately for various reasons each of the projects in these case studies faced different challenges preventing their smooth running. The experience from these projects has provided valuable lessons for future similar projects.

The Ghana water industry

Ghana is a West African country with a total land area of 238,533 square kilometres. In 2010, the total population was estimated to be 24.392 million (World Bank 2011a), where approximately half of the population live in rural areas and the other half in urban areas (Ghana Integrity Initiative 2011). Currently, the population is growing at a rate of 1.5 per cent and 4.5 per cent for the rural and urban areas respectively per annum, indicating a rapid migration from rural to urban areas (WaterAid 2010; 2011). The growing population in urban areas has increased the need for more and better public facilities and services.

The water sector in Ghana was originally managed by the government company Ghana Water Company Limited (GWCL). But with the significant deterioration in water facilities, pure public sector management proved not to be the best solution (Larbi 2012). It was reported that only 56 per cent of the country was covered by water services (Ministry of Water Resources, Works and Housing 2011) and only 40 per cent of the urban population was served (Ghana Water Company Limited 2008; Training Research and Networking for Development 2003). These statistics indicated high water loss, lack of access, poor water quality, low cost recovery, water shortages and poor reliability of supplies (Larbi 2012; Ghana Water Company Limited 2008; Nyarko 2007).

The water demand in Ghana far exceeds the supply. According to the water distribution design standards, 75–150 litres of water is required per head per day (l/hd), whilst in reality the actual consumption ranges between 40–60 l/hd. This range is further lowered in those poorer communities within Ghana. The daily water demand is currently 939,000 cubic metres, but GWCL is only able to supply 551,000 cubic metres/day (Ghana Integrity Initiative 2011). Ghana Water Company Limited (2008) conducted a study which showed that US$1.49 billion would be required to meet demands for water supplies in Ghana up to year 2020.

Studies have reflected that the water supply in many parts of Ghana is highly unreliable (Ghana Integrity Initiative 2011). In some areas consumers may be receiving water for three days a week or fewer, or even no water supply at all in some higher locations. WaterAid's (2011) study showed that in the country's capital only 25 per cent of the population received a water supply twenty-four hours a day, 30 per cent of the population received water for twelve hours a day for five days a week, 35 per cent of the population received water for two days a week and the population surrounding the capital had no water supply at all, which has forced them to use unsafe alternative sources which are also ten times more costly than normal (Larbi 2012).

A high water loss from the facilities has also been experienced due to water leakages and theft. Water losses can be over half the amount produced due to the old and deteriorated pipelines caused by the lack of maintenance and attention (Kauffmann and Perard 2007). The unprotected system has also become a target of theft, where huge amounts are continuously stolen (Ofosu 2004).

The problems of Ghana's water industry are further worsened by the lack of financial resources to provide for the required maintenance, monitoring and expansions. Private financing appears to be an attractive option of the local government to deal with these much needed public utilities.

The need for private sector's involvement

Similar to other PPP projects, the private sector is often involved in public projects in order to provide expertise, efficiency and finance that cannot be provided as well by the government. The private sector's involvement in water projects is also necessary to overcome the challenges that the urban water sector faces. These challenges have been grouped into four main categories by Zaato (2011) namely operational, financial, technological and managerial. These challenges have been summarised in Table 11.1 within each of the four categories.

One of the main challenges faced by the water sector in Ghana is serious financial constraint. This is also a common problem in other developing countries. In Ghana, the water sector is typically financed either externally by international donors or internally via tariffs collected, and government support. Amongst these funding sources, external funding is the main approach adopted. This can be in the form of grants, loans, or a combination, providing over 90 per cent of the funding required. Some of the previous major donors include Danish International Development Agency, Canadian International Development Agency, World Bank, Deutsche Gesellschaft für Technisches Zusammenarbeit,

Table 11.1 Challenges of the urban water sector

Area	Challenges
Operational	Deterioration of distribution network
	Poor maintenance and asset management
	Imbalance of water supply and demand
	Poor labour productivity
	Weak workforce
	Short service hours and unreliable supply
	Low urban service coverage
Financial	Lack of government funding
	Cross-subsidies from government
	Heavy reliance on investors
	Poor billing/tariff collection rate
	Low production of water and sales
	Over-staffing
	High operational expenditure
Technological	Old/faulty machinery
	Water leakage
	Water quality problems
	Low number of connected consumers
	Low metering ratio
Managerial	Poor managerial ability
	Technical and managerial roles mixed
	Low qualification of workforce

Source: Zaato 2011; Ameyaw and Chan 2012a (with permission from Ernest Ameyaw and Emerald Group Publishing Ltd).

Agence France de Development, International Monetary Fund, European Union, and so on. From 1990 to 2003, these major donors alone contributed approximately US$500 million towards water projects in Ghana (African Development Bank and Organisation for Economic Co-operation and Development 2011). This figure reflects the heavy reliance on external funding. Table 11.2 shows the breakdown between external and internal funding for 2006 and 2007 in Ghana (Ghana Integrity Inititative 2011). The figures confirm that during this period, external funding from international donors was the main source of funding for water projects in Ghana.

The limited internal funding was derived from tariffs collected, connection fees and government taxes. Minimal financial support is provided by the local government (Ministry of Water Resources, Works and Housing 2011). Because of the high levels of poverty the small population paying for water supplies is unable to support the operation of these services. As such, external funding became a vital means to develop water facilities in Ghana. Unfortunately, these externally funded projects have been quite disappointing. Some of the problems which arose included poor management, inefficient operation, government interference and weak regulatory and monitoring systems (Ministry of Water Resources, Works and Housing 2011).

Table 11.2 Proportion of external and internal funding for water projects in Ghana during years 2006 and 2007

Type of funding	Location of water supply	2006		2007	
		Million US$	%	Million US$	%
Internal funding (tariffs, fees, government)	Rural	2.38	2.50	3.36	2.90
	Urban	1.70	1.80	2.06	1.80
	Sub-total	4.08	4.30	5.42	4.70
External funding (international donors)	Rural	58.14	60.30	60.27	52.5
	Urban	34.15	35.4	49.22	42.80
	Sub-total	92.29	95.70	109.49	95.30
Total		96.37	100	114.91	100

Source: Ghana Integrity Initiative 2011; Ameyaw and Chan 2012b (with permission from Mr Ernest Ameyaw).

Public–private partnership for Ghana's water industry

Financial support for Ghana's water industry is much needed in order to uplift its overall quality and expand the serviced area. PPP has become highly discussed as an option for uplifting Ghana's water industry (Eguavoen and Youkhana 2008; Fuest and Haffner 2007); both supportive and unsupportive voices have been heard.

Case 1: The 'attempted' lease contract

Background

Since 1995, the Ghanaian government has heavily encouraged private sector involvement in the water industry as a means of improving service levels and leveraging private capital (Larbi 2012). Foreign consultants commissioned by the World Bank deliberated on several PPP models for the water industry (Halcrow and Partners 1995). One of these was the lease option adopted in 1995 after consultations with stakeholders including GWCL staff, donor agencies, government departments and agencies, the private sector, and non-governmental organisations (NGOs) (Fuest and Haffner 2007).

A business framework was prepared for the lease contract approach which emphasised technical, legal and financial issues (Larbi 2012). Two lease packages were opened up for bids in 1999. The first was a ten-year contract for the management of seventy-three water systems and the second was a thirty-year contract for twenty-seven water systems. At the same time Ghana Water and Sewerage Corporation (GWSC) was converted into a publicly owned company, GWCL, with an objective of strengthening its top management. All pre-qualified bidders were large international corporations including Bi-water, Vivendi, Suez and Saur (International Fact-Finding Mission 2011; Integrated Social Development Centre 2011a). The successful bidders would take over the operation and management of the facilities, as well as investing a further US$70 million for rehabilitation, renewal, and improvement of the water systems (Fuest and Haffner 2007). However, the responsibility for securing finance and executing the extension of the water systems remained with the government.

Challenges

Several factors led to the delay and subsequent cancellation of the lease contracts in 2002. The idea of adopting the PPP approach for the water sector attracted opposition from civil society groups who embarked on anti-PPP campaigns. Public resistance stemmed from the fear of water tariff increases, staff dismissals from the GWCL and the notion that vital utilities such as water should not be left at the decision of the private sector (Fuest and Haffner 2007). The large number of redundancies was one of the main triggers leading to the contracts coming to an end. Staff numbers were reduced to half in order to reduce operation costs (Integrated Social Development Centre 2011a) and in turn allow for sufficient cash flows for future investment in infrastructure expansion and maintenance. Further arguments against PPP were that the policy was not proper, and lacked transparency, and also the experience and ability of the winning bidders came under question (Larbi 2012). Public resistance has long been recognised as major risk in the failure of private investment in other countries (Hall *et al.* 2011).

Additional factors that led to the projects coming to a halt include the poor global economic trend at the time causing an unfavourable environment for private investment, and also the regional political instability caused by the outbreak of war in neighbouring Ivory Coast in 2002. The problems together affected the bidders' decision on their involvement and investment amount (Amenga-Etego 2003; Nii Consult 2003). The experience from these lease contracts diverted the government to adopting a management contract instead.

Case 2: The five-year management contract

Background

The management contract was supported by external donor agencies under the Urban Water Project which aimed at upgrading the water supply infrastructure and advancing the PPP policy for the water sector in Ghana (Larbi 2012). By using a management contract, the government hoped to restore GWCL financially and also to make a significant improvement in the commercial operations of the company (Ministry of Water Resources, Works and Housing 2011). The contract was awarded to Aqua Vitens Rand Limited (AVRL), a Dutch–South African joint venture with GWCL in 2005, via a competitive bidding process (World Bank 2011b). The project comprised three key components: (1) network expansion and rehabilitation which focused on increasing the quantity of bulk water for distribution; (2) public–private partnership development which aimed at improving customer satisfaction; and (3) capacity building and project management which focused on training and research for urban water supply reform.

In order to finance this project a number of donors were secured at the start to provide US$120 million in total. This included a credit of US$103 million from the World Bank which was later turned into a grant, US$5million from the Nordic Development Fund, and US$12 million from the government. Over 70 per cent of the funding (US$91.8 million) was allocated for the rehabilitation of existing water treatment and transmission infrastructure, as well as the extension of water networks to serve low-income areas and those not currently served. It is hoped that this will eliminate small-scale private water vendors from selling at unreasonable prices. Furthermore, another US$10 million from the social connection fund was provided to support those low-income consumers to become connected to the GWCL distribution networks (World Bank 2011b; Fuest and Haffner 2007).

The ownership of the asset and the investment of the extension and rehabilitation remained with GWCL whilst the management and operation of the urban water systems was outsourced to the private operator. Specific responsibilities of the private operator include water production, customer billing, revenue collection, operation and maintenance of water systems, and reduction in water loss by 5 per cent each year. The government's intention was that after five years, the improvement in the urban water supply situation would permit AVRL to proceed with a lease-affermage contract.

Challenges

There were several reasons for the project's failure. First, a large number of employees were made redundant. The redundancies were an attempt to improve operational efficiencies in general. Under the management contract GWCL was obliged to restructure the commercial department, develop economic tariffs, develop organisational functions, and introduce robust measures to reduce high water-loss levels and also to downsize the staff by almost 40 per cent (1,600 out of 4,300 workers) (Integrated Social Development Centre 2011b). This strategy is questionable as other regions in Africa such as Ivory Coast and Mali have been able to launch successful PPP water projects without introducing large redundancies as a means to reduce costs (Fall *et al.* 2009; Marin 2009). Furthermore, the World Bank provided US$10 million to pay off the redundant staff in this project (World Bank 2011b), but it is still unclear how and when these staff members were actually compensated (Integrated Social Development Centre 2011b).

Another problem with the project was the commitment and knowledge of the local government. The public sector failed to monitor the project to ensure efficient delivery of the performance targets. The original intention was that an independent regulator named the Public Utilities Regulatory Commission (PURC) would be responsible for examining and approving the service tariffs. But unfortunately, their role did not prove to be effective as they lacked the authority as well as the financial and human resources to enforce their duties (Fuest and Haffner 2007). Furthermore, Fuest and Haffner (2007) criticised the fact that alternative PPP options based on the local context had not been well considered.

The PPP approach adopted may not have been the best option. There are a wide range of options that could have been evaluated to better suit the local conditions and context. This thought was echoed by Fuest and Haffner (2007) who believed that more comparative research would be needed to identify the successful experiences of PPP in order to design and implement a suitable approach for Ghana. Their study, however, did not address the potential challenges and risks involved and their allocation amongst the parties.

Inherent risks were identified as another area causing the project to fail. The project started off with many problems already, including the highly deteriorated water infrastructure and operational difficulties. The inherent risks which contributed to its failure included the shortcomings found in the Memorandum of Understanding between AVRL and GWCL. AVRL was unable to provide competent managerial personnel as indicated initially; there was a weak regulatory framework to monitor AVRL, and lowest performance measures were not set out in the management contract (Kudom-Agyemang 2011). An important lesson learnt from this case is that the responsibilities and risks of a project must be appropriately identified and allocated amongst the parties in an equitable manner at the beginning (Asian Development Bank 2009; World Bank 2011b). Especially in the water sector, many PPP projects are flawed partly due to inappropriate risk allocation (Roger 2009).

Marin (2009) suggested in his study that the performance of the management contract could be assessed in four key areas including: expansion of service coverage; quality of the services provided; operational efficiency of the utility; and tariff levels imposed on consumers. It was observed through this case study that AVRL failed to meet the contractual targets in terms of increasing the general public's access to piped water; reliability of the service; operational efficiencies due to 51 per cent water loss at the end of the contract period; failure to meet water quality standards; affordability; and financial sustainability of service partly due to high energy (electricity) usage (Integrated Social Development Centre 2011b).

Case 3: The PPP management model for small water projects

Management contracts are frequently used for those large-scale or complex water facilities with the participation of international water companies, and commonly referred to as Domestic Private Sector Participation. Nevertheless there has been a rising trend for small water projects to be managed in a similar manner (Water and Sanitation Program 2010). This approach has shown great potential in terms of both increased coverage and revenues (Lazarte 2011). But the management of these small water projects can vary drastically depending on the people involved.

Figure 11.1 illustrates the management model adopted for small water projects in Ghana. There are seven key stakeholders involved in this model: (1) the local government; (2) the Water and Sanitation Development Board (WSDB); (3) the private operator; (4) the Water and Sanitation Committees (WATSAN); (5) the water vendors; (6) the consumers with connection; and (7) consumers who pay as they use.

These management contracts are awarded by the local government through national competitive bidding. The winning private sector bidder will engage in a five-year operation and maintenance agreement with the Water and Sanitation

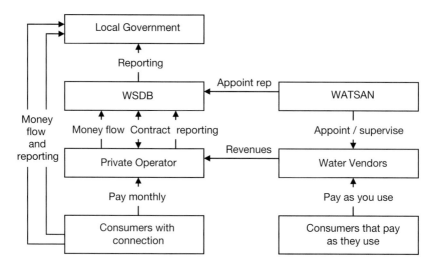

Figure 11.1 The PPP management model for small water projects (Akanbang 2011; Ameyaw and Chan 2012b; Tuffour 2011) (with permission from Ernest Ameyaw)

Development Board (WSDB) acting on behalf of the local government (Tuffour 2011). The local government will remain the legal owner of the water facilities and is therefore responsible for any future expansion, major rehabilitation, technical assistance and capacity building (Tuffour 2011; Maku 2011).

WSDB would be responsible for the operation and management of the water facilities. They will engage a private operator to carry out the actual works required, while they would take up the supervisory role to ensure that the private party is performing according to the required standard and other agreements. They will also communicate with the beneficiary communities via the Water and Sanitation Committee (WATSAN).

The private operator will be re-engaged to operate and manage the daily operations of the water facilities. Their key role is to ensure the production and distribution of safe water, propose appropriate tariffs, collect revenue and conduct maintenance. The private party will report to the local government and WSDB on a quarterly and annually basis regarding the financial and technical performance of the water facilities (Van-Ess 2009; Tuffour 2011).

The WATSAN is responsible for appointing and supervising the water vendors. It will also have representatives within the WSDB.

There are two types of consumers, those that have regular connection and those who pay as they use. For those that have a regular connection with the private operator, they will pay them on a monthly basis. Consumers who pay as they use will take water from public standposts supplied by the water vendors.

The main difference between contracts for small water projects and typical urban water supply projects is that the operator is not paid a fixed fee. Instead the operator is entitled to 75 per cent of the revenue for operational expenses and profit purposes. WSDB is given 10 per cent for ensuring sanitation and hygiene promotion. The remaining 15 per cent is given to the local government as reserves for infrastructure expansion. Furthermore, a unique difference of small water projects is that water vendors keep 20 per cent of their monthly sales from water standposts to cover their expenses and profit.

Challenges

Similar to the other case studies presented in this chapter, this model also experienced some problems. The WSDB itself had many problems; it was not given the legal authority to monitor the operator. In addition, they did not have the authority to approve tariff rates, approval had to be sought from the local government (Fuest et al. 2011). Political interference within the WSDB was also a serious problem (Tuffour 2011). Different WSDBs would vary drastically depending on the ability of the personnel to execute regulatory functions.

Maku (2011) recorded the occurrence of corruption by WATSANs. Problems within the model and also the low profit margins from small-scale projects are main drivers (Laube and van de Giesen 2006). Furthermore, the current approach for splitting the revenue does not account for asset considerations, environmental factors, operating costs, or potential risks.

These projects have been conducted in a much similar way but in fact there are many influencing factors which should be taken into consideration so that necessary changes to the model could be made. For example, the state of the existing water infrastructure, the technology adopted, religious and cultural characteristics of customers, availability of alternative water sources, financial status of customers, amount of government bills and actual water consumption (Berkoh et al. 2004). A case-by-case profitability analysis could be conducted.

Chapter summary

This chapter has presented three case studies adopting different forms of the PPP model for water projects in Ghana. For the 'attempted' lease contract some of the challenges included the anti-PPP campaigns, general public resistance as a result of concern regarding water tariff increases and staff dismissals, the thinking that vital utilities should not be left at the discretion of the private sector, the large redundancy number, the PPP model not being proper and lacked transparency, inadequate experience and inability of the winning bidders, poor global economic trend, regional political instability and the lack of bidders.

The five-year management contract was similar to the 'attempted' lease contract in that a large number of employees were made redundant, other challenges included the lack of commitment and knowledge of the local government, the independent regulator lacking authority, financial and human resources to enforce their duties, other procurement options could have been evaluated to better suit the local conditions and context, inherent risks including the highly deteriorated water infrastructure and operational difficulties, shortcomings found in the Memorandum of Understanding, weak regulatory framework, and that the lowest performance measures were not set out.

Similar to the five-year management contract, the model for small water projects also faced the problem that the regulator was not given the legal authority to monitor the operator. In addition, there was political interference, corruption, low profit margins from small-scale projects, the revenue split did not account for asset considerations, environmental factors, operating costs and potential risks, and also the performance of these projects varied depending on their conditions.

The analyses of these projects showed that there were many underlying challenges with using the PPP approach for water projects in Ghana. Although the original motives for adopting PPP were positive, the analyses have showed that the challenges mentioned in this chapter need to be overcome in order to carry out these projects successfully.

Epilogue

The use of public-private partnership (PPP) for construction projects is definitely not a new approach, instead its history spans a few centuries. Over the years we have seen how the application has extended from the traditional transportation type projects to more complex social projects such as art and culture facilities. The extent that PPP is used in the construction industry has definitely developed in terms of project size, complexity, creativity, innovation and technology.

The financial arrangements have also been repackaged into different forms and types, where more possibilities have been opened up. Traditional PPP projects have very much relied on full private financial support, whereas modern projects have shown that PPP projects can also be supported solely by the public sector. PPP projects for the construction industry have developed in that private financial support is not the only attraction of this model. Of course it is still one of the main drivers for many governments and projects across the world, but there are those projects that may use PPP for its other advantages such as tapping into the private sector's expertise, skills, innovation, management ability, technology and creativity. These benefits may not be possible with sole public sector leadership, as the main and best role of governments across the world is to perform admin-istration effectively. Consequently, PPP projects can also be financed partially or fully by the public sector. Nevertheless, full private financial support is still a key attraction to many governments whose fiscal reserves are not comfortable. In some cases, governments may need to deliver many large projects within a short space of time and hence a looser budget would be more advantageous, or at other times governments may simply want to keep reserves for investment in other areas. As a result, various forms of financial packages are used and supported by different governments worldwide.

Risk sharing, control and allocation have also been the main concerns related with PPP projects in the construction industry. It is reasonable to suggest that the party best able to manage that risk should take up that risk. Nevertheless, this may be easier said than done for many PPP projects. Lessons learnt from previous projects have shown that estimations and predictions may not always be accurate, and in the event that an unexpected scenario occurs the extent of the risks may be too extreme for one party alone to bear. In other situations, risks have been wrongly allocated to the parties and when they do occur, these parties are simply unable to handle them alone. Ideally, the party most able to handle that risk should step in when it occurs, but at the end of the day a PPP project is very much about sharing the risks and their consequences. Although a sole responsible party should be assigned, the other parties should also assist and share the remedial actions and consequences.

PPP projects in the construction industry can also differ greatly when they are led by different governments and in different countries. There is no one model which can be applied to all governments and countries. The experience of other countries is very valuable but experience tells us that adjustments need to be made to apply one model to another country. Different governments and countries across the world will differ in terms of experience, history, mindset, culture, priorities, financial security, and stability. With these differences in consideration, no PPP model is exactly the same. Hence there has been an evolution of different forms and types of PPP models across the world. Nevertheless, we believe that PPP projects can be in any form or type where the public and private parties join forces to deliver a public facility and or service for the benefit and need of the general public.

Finally we hope that our readers have found this book an interesting source documenting the evolution of PPP for the construction industry worldwide. We are confident that PPP will continue to play an important role in delivering construction projects for governments worldwide where value for money can be achieved. In addition, we believe that it will continue to change and develop to suit the timely needs within our industry.

References

AAP General News Wire (2006a) 'War of words erupts again between Lemma and tunnel boss', 4 August 2006.

AAP General News Wire (2006b) 'Cross city not viable, higher prices not the answer', 26 August 2006.

AAP General News Wire (2006c) 'Lane Cove Tunnel Road changes may be as bad as cross city', 21 August 2006.

AAP General News Wire (2006d) 'Motorists have right to be angry over toll inequities', 20 September 2006.

AAP General News Wire (2006e) 'Cross city boss says Lemma fails to show leadership', 4 August 2006.

AAP General News Wire (2006f) 'Tunnel operators seek millions in compensation for changes', 26 August 2006.

Abdul-Rashid, A. A., Puteri, S. J. K., Ahmed, U. A. and Mastura, J. (2006) 'Public private partnerships (PPP) in housing development: the experience of IJM Malaysia in Hyderabad, India', Proceedings of the Conference *Accelerating Excellence in the Built Environment*, 2–4 October 2006, Birmingham, UK.

Adams, J., Young, A. and Wu, Z. (2006) 'Public private partnerships in China – system, constraints and future prospects', *International Journal of Public Sector Management*, 19(4), 384–96.

African Development Bank and Organisation for Economic Co-operation and Development (2011) *African Economic Outlook*, http://www.oecd.org/dataoecd/26/51/38562673.pdf (accessed on 12 November 2011).

Ahadzi, M. and Bowles, G. (2004) 'Public-private partnerships and contract negotiations: an empirical study', *Construction Management and Economics*, 22(9), 967–78.

Akanbang, B. A. A. (2011) 'Public–private partnerships in small town water supply in Tumu', Tripartite Partnership Project, Ghana, www.ghana.watsan.net/page/777 (accessed on 11 December 2011).

Akbiyikli, R. and Eaton, D. (2004) 'Risk management in PFI procurement: a holistic approach', *Proceedings of the 20th Annual Association of Researchers in Construction Management (ARCOM) Conference*, Heriot-Watt University, Edinburgh, UK, 1–3 September 2004, 1269–79.

Akintoye, A., Beck, M., Hardcastle, C., Chinyio, E. and Asenova, D. (2001) 'The financial structure of private finance initiative projects', *Proceedings of the 17th ARCOM Annual Conference*, Salford University, Manchester, UK, 1361–9.

Akintoye, A., Beck, M. and Hardcastle, C. (2003) 'Public-Private Partnerships: managing Beijing Municipal Commission of Development and Reform', *Proceedings of International Forum on Infrastructure Marketization*, http://www.bjpc.gov.cn/zt/sheshi (accessed on 18 March 2008).

Al-Sharif, F. and Kaka, A. (2004) 'PFI/PPP topic coverage in construction journals', *Proceedings of the 20th Annual ARCOM Conference*, 1–3 September 2004, Heriot-Watt University, UK, (1), 711–19.

Amenga-Etego, R. (2003) 'From social contract to private contracts: the privatisation of health, education and basic infrastructure', *Public Agenda*, Accra, 25 August 2003.

Ameyaw, E. E. and Chan, A. P. C. (2012a) 'Assessing the performance of the urban water utility under a management contract – a case from Ghana', *International Journal of Professional Management*, 6(6), June 2012.

Ameyaw, E. E. and Chan, A. P. C. (2012b) 'The private sector's involvement in the water industry of Ghana', *Journal of Engineering, Design and Technology*, forthcoming.

Antiquities and Monuments Office (2010) *Definition of the Gradings of Historic Buildings*, http://www.amo.gov.hk/en/built3.php (accessed on 13 May 2010).

Antiquities and Monuments Office (2012) *Introduction to 1444 Historic Buildings*, http://www.amo.gov.hk/form/AAB_brief_info_en.pdf (14 February 2012).

Apple Daily (2008) 明智決定謝絕財團免加價冇王管 (Chinese version only) English translation 'Wise decision to prevent consortia from raising tolls', *Apple Daily*, 6 August 2008, Hong Kong.

Asian Development Bank (2009) *Public-Private Partnership Handbook*, Asian Development Bank, Manila.

Askar, M. M. and Gab-Allah, A. A. (2002) 'Problems facing parties involved in build, operate, and transport projects in Egypt', *Journal of Management in Engineering*, 18(4), 173–8.

Awodele, O. A., Ogunlana, S. O. and Akinradewo, O. F. (2012) 'Evaluation of public private partnership (PPP) as alternative procurement route for infrastructure development: case of Nigeria megacity', in Laryea, S., Agyepong, S. A., Leiringer, R. and Hughes, W. (eds) *Proceedings of the 4th West Africa Built Environment Research (WABER) Conference*, 24–26 July 2012, Abuja, Nigeria, 329–44.

Berg, S. V., Pollitt, M. G. and Tsuji, M. (2002) *Private Initiatives in Infrastructure*, Edward Elgar, Cheltenham, UK.

Berkoh, C., Hirsch, D., Larbi, E., Simavi, F. L., Ntow, S., Smet, J. and Simavi, M. T. (2004) 'Report on the feasibility of public-private partnership for sustainable water supply to the urban poor in Ghana', unpublished.

Birnie, J. (1999) 'Private Finance Initiative (PFI) – UK construction industry response', *Journal of Construction Procurement*, 5(1), 5–14.

Borzel, T. A. and Risse, T. (2005) 'Public private partnerships: effective and legitimate tools of transnational governance?' in Grande, E. and Pauly, L. (eds) *Complex Sovereignty: Reconstituting Political Authority in the Twenty-First Century*, University of Toronto Press.

Boussabaine, A. (2007) *Cost planning of PFI and PPP building projects*, Taylor & Francis.

Bovaird, T. (2004) 'Public–private partnerships: from contested concepts to prevalent practice', *International Review of Administrative Sciences*, 70(2), 199–215.

Braadbaart, O., Zhang, M. and Wang, Y. (2008) 'Managing urban wastewater in China: a survey of build-operate-transfer contracts', *Water and Environment Journal*, 23, 46–51.

British Columbia (1999) *Public Private Partnership – A Guide for Local Government Ministry of Municipal Affairs*, British Columbia Government.

Brown, G. (2008) 'We're stuck on the bridge of sighs', *The Standard* (Hong Kong), 8 August 2008.

Bult-Spiering, M. and Dewulf, G. (2006) *Strategic Issues in Public-Private Partnerships: An International Perspective*, Williston, VT, USA: Wiley-Blackwell.

Burger, P. (2012) 'The dedicated PPP unit of the South African national treasury, University of the Free State South Africa', http://www.oecd.org/mena/governance/37146964.pdf (accessed on 20 November 2012).

Canadian Council for Public Private Partnerships (2012) *Public Private Partnerships: A Guide for Municipalities*, November 2011, http://www.p3canada.ca/_files/file/P3%20 Guide%20for%20Municipalities%20-%20Engliish%20-%20Final.pdf (accessed on 21 November 2012).

Carrillo, P., Robinson, H., Foale, P., Anumba, C. and Bouchlaghem, D. (2007) 'Participation, barriers and opportunities in PFI: the United Kingdom experience', *Journal of Management in Engineering*, ASCE, 24(3), 138–45.

Chan, D. W. M., Chan, A. P. C. and Lam, P. T. I. (2006) 'A feasibility study of the implementation of public private partnership (PPP) in Hong Kong', *Proceedings of the CIB W89 International Conference on Building Education and Research*, April 10–13, 2006 (under Sub-theme 2.6 – Procurement Management).

Chan, A. P. C., Lam, P. T. I., Chan, D. W. M., Chiang, Y. H., Tang, B. S. and Cheung, E. (2007a) *Component A Report – Public Private Partnership Manual for China Urban Rail Project Development*, Consultancy Report for MTRC Ltd, commissioned by the Asian Development Bank, TA4724-PRC: Application of Public Private Partnership in Urban Rail-based Transportation Project, July 2007.

Chan, A. P. C., Sidwell, T., Kajewski, S., Lam, P. T. I., Chan, D. W. M. and Cheung, E. (2007b) 'From BOT to PPP – a Hong Kong example', *Proceedings of the 2007 International Conference on Concession Public/Infrastructural Projects (ICCPIP)*, Dalian University of Technology, China, 24–26 August 2007, 9:010–018.

Chan, A. P. C., Kajewski, S. and Cheung, E. (2008a) 'Obstacles in adopting PPP – lessons from Australia', *Projects and Profits*, Special Issue: Managing Construction Projects, The ICFAI University Press, 60, 28–31.

Chan, A. P. C., Lam, P. T. I., Chan, D. W. M. and Cheung, E. (2008b) 'A mechanism for risk sharing in PPP projects – the Sydney Cross City Tunnel case study', *Surveying and Built Environment* (Journal of the Hong Kong Institute of Surveyors), 19(1), 67–80.

Chan, A. P. C., Lam, P. T. I., Chan, D. W. M. and Cheung, E. (2008c) 'Application of public private partnership (PPP) in Hong Kong Special Administrative Region – the critics' perspectives', *Proceedings of the International Conference on Construction in Developing Countries – Advancing and Integrating Construction, Education, Research and Practices* (ICCIDC-I), Pakistan, 4–5 August 2008, 302–11.

Chan, A. P. C., Lam, P. T. I., Chan, D. W. M. and Cheung, E. (2008d) 'Application of public private partnership (PPP) model in procuring infrastructure projects in Hong Kong', *Proceedings of the CIB W055–W065 Joint International Symposium: Transformation Through Construction*, Dubai, 17–19 November 2008, 190–1.

Chan, A. P. C., Lam, P. T. I., Chan, D. W. M., and Cheung, E. (2008e) 'Application of public private partnership (PPP) in Hong Kong Special Administrative Region – the private sector's perspective', *Proceedings of the Chinese Research Institute of Construction Management (CRIOCM) 2008 International Symposium on Advancement of Construction Management and Real Estate*, Beijing, China, 31 October–3 November 2008.

Chan, A. P. C., Lam, P. T. I., Chan, D. W. M., Cheung, E. and Ke, Y. J. (2008f) 'Measures that enhance the achievement of value-for-money in PPP projects', *Proceedings of the Building Abroad Conference – Procurement of construction and reconstruction projects in the international context*, IF Research Group, i-Rec and CIB, Université de Montréal, Canada, 23–25 October 2008.

Chan, A. P. C., Lam, P. T. I., Chan, D. W. M., Cheung, E. and Ke, Y. J. (2009) 'Drivers for adopting PPP – a comparison between China and Hong Kong Special Administrative Region', *Journal of Construction Engineering and Management*, ASCE, 135(11), 1115–24.

Chan, A. P. C., Lam, P. T. I., Chan, D. W. M., Cheung, E. and Ke, Y. J. (2010) 'Critical success factors for public private partnerships (PPP) in infrastructure developments: a

Chinese perspective', *Journal of Construction Engineering and Management*, ASCE, 136(5), 484–94.

Chege, L. W. (2001) 'Private financing of construction projects and procurement systems: an integrated approach', *Proceedings of the CIB World Building Congress*, April 2001, Wellington, New Zealand.

Chen, B. (2008) 'Bridge set for early finish on new deal', *The Standard* (Hong Kong), 4 August 2008.

Chen, B. and Lee, D. (2008) 'Bridging the divide', *The Standard* (Hong Kong), 6 August 2012.

Chen, C. and Doloi, H. (2008) 'BOT application in China: driving and impeding factors', *International Journal of Project Management*, 26, 388–98.

Chen, M. and Shi, X. J. (2009) 'Risk allocation of public-private partnerships in public stadium construction projects', *Proceedings of CRIOCM 2009 International Symposium on Advancement of Construction Management and Real Estate*, 29–31 October 2009, Nanjing, China,1(3), 338–42.

Cheung, E. and Chan, A. P. C. (2009a) 'Is BOT the best financing model to procure infrastructure projects? A case study of the Hong Kong–Zhuhai–Macau bridge', *Journal of Property Investment and Finance*, 27(3), 290–302.

Cheung, E. and Chan, A. P. C. (2009b) 'Suitability of procuring large public works by public private partnership (PPP) - a case study of the Kai Tak cruise terminal project in Hong Kong', *Proceedings of the 34th Australasian Universities Building Education Association (AUBEA) Annual Conference*, Adelaide, Australia, 7–10 July 2009, Paper 31.

Cheung, E. and Chan, A. P. C. (2011a) 'An evaluation model for assessing the suitability of public private partnership (PPP) projects', *Journal of Management in Engineering*, ASCE, 27(2), 80–9.

Cheung, E. and Chan, A. P. C. (2011b) 'Risk factors of public private partnership projects in China: a comparison between the water, power and transportation sectors', *Journal of Urban Planning and Development*, ASCE, 137(4), 409–15.

Cheung, E. and Chan, A. P. C. (2012) 'Revitalising historic buildings through partnership scheme: a case study of the Mei Ho House in Hong Kong', *Property Management*, 30(2).

Cheung, E., Chan, A. P. C. and Kajewski, S. (2009a) 'Reasons for implementing public private partnership projects: perspectives from Hong Kong, Australian and British practitioners', *Journal of Property Investment and Finance*, 27(1), 81–95.

Cheung, E., Chan, A. P. C. and Kajewski, S. (2009b) 'Enhancing value for money in public private partnership projects - findings from a survey conducted in Hong Kong, Australia and the United Kingdom',*Journal of Financial Management of Property and Construction*, 14(1), 7–20.

Cheung, E., Chan, A. P. C. and Kajewski, S. (2010a) 'The public sector's perspective on procuring public works projects; comparing the views of practitioners in Hong Kong and Australia', *Journal of Civil Engineering and Management*, 16(1).

Cheung, E., Chan, A. P. C. and Kajewski, S. (2010b) 'The researcher's perspective on procuring public works projects - findings from Hong Kong and Australian interviewees', *Structural Survey*, 28(4).

Cheung, E., Chan, A. P. C. and Kajewski, S. (2010c) 'Suitability of procuring large public works by PPP', *Engineering, Construction and Architectural Management*, 17(3), 292–308.

Cheung, E., Chan, A. P. C. and Kajewski, S. (2012) 'Factors contributing to successful public private partnership projects: comparing Hong Kong with Australia and the United Kingdom', *Journal of Facilities Management*, 10(1), 45–58.

Civic Exchange, APCO Asia Ltd and Hawker Britton (2005) *Getting PPP Right: Using West Kowloon Cultural District as a Case Study*, Hong Kong.

Cobb, J. M. (2005) 'Financing Bangkok's mass transit', IDC TransGate. http://www.idcworld.com/bangkok.htm (accessed on 23 May 2007).

Commissioner for Heritage's Office (2010a) *Revitalising Historic Buildings Through Partnership Scheme – Mei Ho House: Resource Kits*, Development Bureau http://www.heritage.gov.hk/en/doc/Resource%20kit_Mei_Ho_House.pdf (accessed on 13 May 2010).

Commissioner for Heritage's Office (2010b) *Batch I of the Revitalisation Scheme: Result of Selection*, Development Bureau http://www.heritage.gov.hk/en/rhbtp/ProgressResult_Mei_Ho_House.htm (accessed on 13 May 2010).

Commissioner for Heritage's Office (2010c) *Batch I of the Revitalisation Scheme – Mei Ho House*, Development Bureau http://www.heritage.gov.hk/en/rhbtp/febs.htm?bsIDE3 (accessed on 13 May 2010).

Commissioner for Heritage's Office (2010d) *Batch I of the Revitalisation Scheme: Progress of Processing Applications*, Development Bureau http://www.heritage.gov.hk/en/rhbtp/ProgressApp.htm (accessed on 13 May 2010).

Commissioner for Heritage's Office (2010e) Front View of Mei Ho House, Development Bureau http://www.heritage.gov.hk/images/rhbtp/result/Mei_Ho_House/MHH_1a.jpg (accessed on 26 May 2010).

Commissioner for Heritage's Office (2010f) Original Room of Mei Ho House, Development Bureau http://www.heritage.gov.hk/images/rhbtp/result/Mei_Ho_House/MHH_2a.jpg (accessed on 26 May 2010).

Commissioner for Heritage's Office (2012) *Revitalising Historic Buildings Through Partnership Scheme – North Kowloon Magistracy Resource Kit*, Development Bureau http://www.heritage.gov.hk/en/doc/Resource%20kit_North_Kowloon_Magistracy.pdf (29 February 2012).

Corbett, P. and Smith, R. (2006) 'An analysis of the success of the Private Finance Initiative as the Government's preferred procurement route', *Proceedings of the Accelerating Excellence in the Built Environment Conference*, Birmingham, UK, 2–4 October 2006.

Cratchley, D. and Jean, P. (2006a) 'Govt may compensate Lane Cove Tunnel operators', AAP General News Wire, 28 August 2006.

Cratchley, D. and Jean, P. (2006b) 'State government may compensate Lane Cove Tunnel owners', AAP General News Wire, 28 August 2006.

Cross City Tunnel Proprietary Ltd (2007) *Cross City*, www.crosscity.com.au (accessed on 30 May 2007)

Detail Commercial Solicitors (2012) *Nigeria PPP Review*, http://www.detailsolicitors.com/media/archive2/articles/PPPreview.pdf (accessed on 20 November 2012).

Development Bureau (2012) *About the Scheme*, http://www.heritage.gov.hk/en/rhbtp/about.htm (accessed on 28 December 2012).

Dhaene, G. (2008) 'PPP initiatives and the HR crisis in the health sector', *Proceedings of the Global Health Workforce Alliance Conference*, March 2008, Kampala, Uganda.

Diário Oficial da República Federativa do Brasil (2004) *Brazilian Federal Law 11079*, 30 December 2004, Brasília, DF, 21 December 2004.

Drew, J. (2005) 'Public private partnerships – opportunities and challenges', *Proceedings of the Conference on Public Private Partnerships – Opportunities and Challenges*, 22 February 2005, Hong Kong.

Duffield, C. F. (2001) 'An evaluation framework for privately funded infrastructure projects in Australia', unpublished PhD thesis, University of Melbourne, November 2001.

Eaton, D., Akbiyukli, R. and Dickinson, M. (2006) 'An evaluation of the stimulants and impediments to innovation within PFI/PPP projects', *Construction Innovation*, 6, 63–77.

Efficiency Unit (2001) *Serving the Community by Using the Private Sector*, June 2001, Hong Kong Special Administrative Region.

Efficiency Unit (2002) *Project 2002 – Enhancing the Quality of Education in Glasgow City Schools by Public Private Partnership*, Hong Kong Special Administrative Region Government.

Efficiency Unit (2003) *Serving the Community by Using the Private Sector – An Introduction Guide to Public Private Partnerships (PPPs)*, August 2003, Hong Kong Special Administrative Region.

Efficiency Unit (2007) *Serving the Community by Using the Private Sector Policy and Practice* (second edition), January 2007, Hong Kong Special Administrative Region Government.

Efficiency Unit (2008) *Serving the Community by Using the Private Sector – An Introduction Guide to Public Private Partnerships (PPPs)* (second edition), March 2008, Hong Kong Special Administrative Region.

Efficiency Unit (2012a) *Public Private Partnership Overview*, http://www.eu.gov.hk/english/psi/psi_ppp/psi_ppp_over/psi_ppp_over.html (accessed on 6 August 2012).

Efficiency Unit (2012b) *Case Summary: University College London Hospital (UCLH) Redevelopment – Improving the Standard of Healthcare by Public Private Partnership* http://www.eu.gov.hk/english/psi/psi_ppp/psi_ppp_cases/files/uclh_redevelopment.pdf (accessed on 6 August 2012).

Efficiency Unit (2012c) *Policy and Practical Guides to PSI, Outsourcing and PPPs*, http://www.eu.gov.hk/english/psi/psi_guides/psi_guides_ppgpop/psi_guides_ppgpop.html (accessed on 28 December 2012), Hong Kong Special Administrative Region Government.

Eguavoen, I. and Youkhana, E. (2008) 'Small towns face big challenge: the management of piped systems after the sector reform', GLOWA Volta Project working paper, University of Bonn, Germany, 2008.

El-Gohary, N. M., Osman, H. and El-Diraby, T. E. (2006) 'Stakeholder management for public private partnerships', *International Journal of Project Management*, 24(7), 595–604.

English, L. M. and Guthrie, J. (2003) 'Driving privately financed projects in Australia: what makes them tick', *Accounting, Auditing and Accountability Journal*, 16(3), 493–511.

Entwistle, T. and Martin, S. (2005) 'From competition to collaboration in public service delivery: a new agenda for research', *Public Administration*, 83(1), 233–42.

Environment, Transport and Works Bureau (2004) *Reference Guide on Selection of Procurement Approach and Project Delivery Techniques*, Technical Circular (Works) No. 32/2004, October 2004, Hong Kong Special Administrative Region Government.

Ernst & Young (2005) *Australian PPP Survey – Issues Facing the Australian PPP Market*, Ernst & Young, November 2005.

European Commission Directorate (2003) *Guidelines for Successful Public-Private Partnerships – Version 1, Directorate-General Regional Policy*, European Commission, February 2003.

Fall, M., Marin, P., Locussol, A. and Verspyck, R. (2009) *Reforming Urban Water Utilities in Western and Central Africa: Experiences with Public-Private Partnerships – Volume 1: Impact and Lessons Learned*, Discussion Paper Series 13, World Bank, Washington DC.

Farrah, T. (2007) 'Brumby wins battle to keep EastLink costs secret', *The Age*, 14 February 2007. http://www.theage.com.au/news/national/brumby-wins-battle-to-keep-eastlink-costs-secret/2007/02/13/1171128974031.html (accessed on 15 May 2007).

Field, K. (2006a) 'Childish act shows NSW not open for business', AAP General News Wire, 10 October 2006.

Field, K. (2006b) 'Sydney's cross city tunnel operators to pursue toll cheats', AAP General News Wire, 20 September 2006.

Flickr (2010) 美荷樓：第一型徙置大廈, 樓高六層, 中間為廁所+浴室 (Chinese version only) English translation 'Mei Ho House: first type of resettlement building, six-storey high, toilet and bathrooms in middle', available at: http://www.flickr.com/photos/milmil302/3734323023/ (accessed on 26 May 2010).

Forum Sara (2008) Western Harbour Crossing Tunnel, http://site.sara.free.fr/photos/HKG-West_Kowloon_Highway-001-Western_Harbour_Crossing_Tunnel-JRL.jpg (accessed on 16 November 2008).

Foster and Partners (2013) Foster and Partners/Dbox, https://fosterandpartners1.box.com/shared/1a7yo6f596#/s/1a7yo6f596/1/119540015/5410204810/1 (accessed on 17 January 2013).

Fuest, V. and Haffner, S. A. (2007) 'PPP – policies, practices and problems in Ghana's urban water supply', *Water Policy*, 9, 169–92.

Fuest, V., Ampomah, B., Haffner, S. A. and Tweneboah, E. (2011) *Mapping the Water Sector of Ghana: An Inventory of Institutions and Actors*, The GLOVA Volta Project, www.glova-volta.de (accessed on 23 December 2011).

Gentry, B. S. and Fernandez, L. O. (1997) 'Evolving public-private partnerships: general themes and urban water examples', *Proceedings of the OECD Workshop on Globalization and the Environment: Perspectives from OECD and Dynamic Non-Member Economies*, Paris, 13–14 November 1997, 19–25.

Ghana Integrity Initiative (2011) *Ghana's National Water Supply Integrity Study*, www.tighana.org/giipages/publication/TISDA%20LAUNCH%20REPORT%202011.pdf (accessed on 10 October 2011).

Ghana Water Company Ltd (2008) *Mid-year review: January to June 2007*, Corporate Planning and ICT Department, Ghana.

Ghobadian, A., Gallear, D., O'Regan, N. and Howard, V. (2004) *Future of the Public Private Partnership, Public Private Partnerships: Policy and Experience*, Palgrave Macmillan, Basingstoke.

Glaeser, E. L. (2001) 'Public ownership in the American city', *National Bureau of Economic Research* http://www.nber.org/papers/w8613.pdf?new_window=1, 7 August 2012.

Grant, T (1996) 'Keys to successful public-private partnerships', *Canadian Business Review*, 23(3), 27–8.

Grimsey, D. and Lewis, M. K. (2004) *Public Private Partnerships: The Worldwide Revolution in Infrastructure Provision and Project Finance*, Edward Elgar, Cheltenham.

Grimsey, D. and Lewis, M. K. (2005) 'Are public private partnerships value for money? Evaluating alternative approaches and comparing academic and practitioner views', *Accounting Forum*, 29(4), 345–78.

Guasch, J. L. (2012) *Granting and Renegotiating Infrastructure Concessions: Doing It Right*, World Bank Institute, Washington, http://www.google.com.hk/books?hl=zh-TW&lr=&id=xBqtbaBM-Z0C&oi=fnd&pg=PR7&dq=concessions&ots=RwSHnQJ48x&sig=huCnKObFQegodLAEC5Lqe3xuHw8&redir_esc=y# (accessed on 26 November 2012).

Gunnigan, L. and Eaton, D. (2006) 'Addressing the challenges that are emerging in the continued increase in PPP use in the Republic of Ireland', *Proceedings of the CIB W89 International Conference on Building Education and Research*, CIB, 10–13 April 2006, Hong Kong.

Halcrow, S. W. and Partners (1995) *Consultancy Services for the Restructuring of the Water Sector*, Final report, Swindon, Wiltshire, Ministry of Works and Housing, Accra.

Hall, D., Lobina, E. and de la Motte, R. (2011) 'Public Resistance to Privatisation in Water and Energy', *Development in Practice*, 15(3,4), www.psiru.org/reports/2005-06-W-E-resist.pdf (accessed on 10 October 2011).

Haslam, R. (2003) 'Focus groups in health and safety research', in Langford, J. and McDonagh, D., *Focus Groups*, Taylor & Francis, UK.

Heald, D. (2003) 'Value for money tests and accounting treatment in PFI schemes', *Accounting, Auditing and Accountability Journal*, 16(3), 342–71.

Hill, C. W. L. (2005) *International Business: Competition in the Global Marketplace*, http://highered.mcgraw-hill.com/sites/0072873957/student_view0/additional_cases.html (accessed on 28 January 2007).

HM Treasury (2003) *PFI: Meeting the Investment Challenge*, July 2003, UK.

HM Treasury (2012) *A New Approach to Public Private Partnerships*, http://www.hm-treasury.gov.uk/d/infrastructure_new_approach_to_public_private_parnerships_051212.pdf (accessed on 27 December 2012).

Ho, R. C. T. (2005) 'How can we capitalize on the concept of PPP?', *Proceedings of the Conference on Public Private Partnerships – Opportunities and Challenges*, Hong Kong, 22 February 2005.

Hodge, G. A. (2004) 'The risky business of public private partnerships', *Australian Journal of Public Administration*, 63(4), 37–49.

Hong Kong Commercial Newspaper (2010) 變身旅舍保健中心 2年後啟用 2.3億活化美荷樓雷生春, (Chinese version only) English translation 'Transformation to hostel and health centre, in use after 2 years, 0.23 billion to revitalise Mei Ho House and Lui Seng Chun', *Hong Kong Commercial Newspaper*, 21 April 2010, available at: http://www.hkcd.com.hk/content/2010–04/21/content_2512252.htm (accessed on 17 May 2010).

Hong Kong Housing Authority (2010) *History of Estates*, available at: http://www.housingauthority.gov.hk/hdw/en/aboutus/events/community/heritage/about.html (accessed on 26 May 2010).

Hong Kong Special Administrative Region Government. (2008) 'Secretary of Transport and Housing Bureau Speaks of the Hong Kong-Zhuhai-Macau Bridge (questions and answers)', Transport and Housing Bureau, Hong Kong (in Chinese) http://www.thb.gov.hk/tc/psp/pressreleases/transport/land/2008/200802293.htm (12 August 2008).

Hong Kong Youth Hostels Association (2010) *Revitalising Historic Buildings Through Partnership Scheme – Mei Ho House as a City Hostel*, 17 February 2009, http://www.heritage.gov.hk/doc/rhbtp/Mei%20Ho%20House.pdf (accessed on 13 May 2010).

Hood, J. and McGravey, N. (2002) 'Managing the risk of public-private partnerships in Scottish local government', *Policy Studies*, 23(1): 21–35.

Howes, R. and Robinson, H. (2005) *Infrastructure for the Built Environment: Global Procurement Strategies* (first edition), Oxford: Butterworth-Heinemann.

Hughes, W., Hillebrandt, P., Lingard, H. and Greenwood, D. (2001) 'The impact of market and supply configurations on the costs of tendering in the construction industry', *Proceedings of the CIB World Building Congress*, April 2001.

Hung, W. T. (2008) *Hong Kong-Zhuhai-Macau Bridge – Transport and Environmental Concerns*, http://www.chamber.org.hk/streaming/ppt/4_HKZMB_wthung.files/frame.htm (accessed on 13 August 2008).

Ibrahim, A. D., Price, A. D. F. and Dainty, A. R. J. (2006) 'The analysis and allocation of risks in public private partnerships in infrastructure projects in Nigeria', *Journal of Financial Management of Property and Construction*, 11(3), 149–64.

Information Services Department (2008a) 'Shatin-Central link construction set for 2010', Hong Kong Special Administrative Region Government, http://www.news.gov.hk/en/category/infrastructureandlogistics/080311/html/080311en06003.htm (accessed 2 March 2008).

Information Services Department (2008b) 'Pact reached on funding Pearl River bridge', Hong Kong Special Administrative Region Government, 5 August 2008, http://www.news.gov.hk/en/category/infrastructureandlogistics/080805/print/080805en06002.htm (accessed on12 August 2008).

Infranews (2008) 'Australian PPPs ride the storm', http://www.nzcid.org.nz/downloads/Australian%20PPPs%20ride%20the%20storm%20-%20Infranews%20Aug%2006.pdf (accessed on 10 September 2008).

Infrastructure Implementation Group (2005) *Review of Future Provision of Motorways in NSW*, New South Wales Government, Australia.

Ingall, P. (1997) *London Underground's Connect Project*, Institution of Electrical Engineers, http://ieeexplore.ieee.org/iel4/5456/14721/00668019.pdf?arnumber=668019 (accessed on 1 May 2007).

Integrated Social Development Centre (2011a) *Privatisation of Ghana's Water*, International Development Select Committee on the World Bank, www.publications.parliament.uk (accessed on 16 November 2011).

Integrated Social Development Centre (2011b) Press Release, www.isodec.org.gh (accessed on 16 November 2011).

International Fact-Finding Mission (2011) *Report of the International Fact-Finding Mission on Water Sector in Ghana*, www.southernlinks.org/pdf/complete_ghana_water.pdf (accessed on 10 October 2011).

Iyer, K. C. and Sagheer, M. (2010) 'Hierarchical structuring of PPP risks using interpretative structural modeling', *Journal of Construction Engineering and Management*, ASCE, 136(2), 151-9.

Jamali, D. (2004), 'Success and failure mechanisms of public private partnerships (PPPs) in developing countries: insights from the Lebanese context', *International Journal of Public Sector Management*, 17(5), 414-30.

Jean, P. (2006) 'Cronulla riot, tunnel were my toughest days: Lemma', AAP General News Wire, 3 August 2006.

Jefferies, M. (2006) 'Critical success factors of public private sector partnershipsA a case study of the Sydney SuperDome', *Engineering, Construction and Architectural Management*, 13(5), 451-62.

Jefferies, M., Gameson, R. and Rowlinson, S. (2002) 'Critical success factors of the BOOT procurement system: reflections from the Stadium Australia case study', *Engineering Construction and Architectural Management*, 9(4), 352-61.

Jin, X. H. (2010) 'Determinants of efficient risk allocation in privately financed public infrastructure projects in Australia', *Journal of Construction Engineering and Management*, ASCE, 136(2), 138-50.

Jin, X. H. (2011) 'A model for efficient risk allocation in privately financed public infrastructure projects using neuro-fuzzy techniques', *Journal of Construction Engineering and Management*, ASCE, http://ascelibrary.org/coo/resource/3/jcemxx/258 (18 March 2011).

Jin, X. H. and Doloi, H. (2008) 'Interpreting risk allocation mechanism in public–private partnership projects: an empirical study in a transaction cost economics perspective', *Construction Management and Economics*, 26, 707-21.

Judd, T. (2008) 'Altair with three roads', http://www.flickr.com/photos/30246562@N05/2994666472/ (accessed on 17 November 2008).

Judiciary of the Hong Kong Special Administrative Region (2012) *Hong Kong Judiciary Annual Report 2004*, http://www.judiciary.gov.hk/en/publications/annu_rept_2004.htm (14 February 2012).

Kanakoudis, V., Papotis, A., Sanopoulos, A. and Gkoutzios, V. (2007) 'Crucial parameters for PPP projects successful planning and implementation', in Schrenk, M., Popovich, V. V., Benedikt, J. (eds) *REAL CORP 007 Proceedings*, Vienna, 20-23 May 2007, 167-84.

Kanter, R. M. (1999) 'From spare change to real change', *Harvard Business Review*, 77(2), 122-32.

Kappeler, A. and Nemoz, M. (2012) *Public Private Partnerships in Europe – Before and During the Recent Financial Crisis*, European Investment Bank, Economic and Financial Report 2010/04, July 2010, http://www.bei.europa.eu/attachments/efs/efr_2010_v04_en.pdf (accessed on 13 November 2012).

Kauffmann, C. and Perard, E. (2007) 'Stocktaking of the water and sanitation sector and private sector involvement in selected African countries', Background note for the regional roundtable on Strengthening Investment Climate Assessment and Reform in NEPAD Countries, Lusaka, Zambia, 27–28 November, 2007.

Ke, Y., Wang, S., Chan, A. P. C. and Cheung, E. (2009) 'Research trend of public-private partnership in construction journals', *Journal of Construction Engineering and Management*, ASCE, 135(10), 1076–86.

Ke, Y., Wang, S. and Chan, A. P. C. (2010) 'Risk allocation in public-private partnership infrastructure projects: a comparative study', *Journal of Infrastructure Systems*, ASCE, 16(4), 343–51.

Khang, D. G. (1998) *Hopewell's Bangkok Elevated Transport System (BETS)*, School of Management, AIT, Bangkok, Thailand.

Khasnabis, S., Dhingra, S. L., Mishra, S. and Safi, C. (2010) 'Mechanisms for transportation infrastructure investment in developing countries', *Journal of Urban Planning and Development*, ASCE, 136(1), 94–103.

Koppenjan, J. F. M. (2005) 'The formation of public-private partnerships: lessons from nine transport infrastructure projects in the Netherlands', *Public Administration*, 83(1), 135–57.

Krueger, R. A. and King, J. A. (1998) *Focus Group Kit 5 – Involving Community Members in Focus Groups*, Sage Publications, US.

Kudom-Agyemang, A. (2011) 'Transitional arrangement favoured after AVR's contract', *Public Agenda*, May 2011, www.allAfrica.com (accessed on November 2011).

Kwan, J. (2005) 'Public private partnerships: public private dialogue', *Proceedings of the Conference on Public Private Partnerships – Opportunities and Challenges*, 22 February 2005, Hong Kong.

Kwok, W. (2009) *Timeline*, http://jmsc.hku.hk/jmsc6030/bridgestory/dossier/timeline/index.html# (accessed on 17 March 2009).

Lam, A. and Chan, Q. (2008) 'Role in bridge plan ends', *South China Morning Post*, 9 August 2008, Hong Kong.

Lam, A. and Lai, C. (2008) 'Beijing cash puts bridge a step closer', *South China Morning Post*, 6 August 2008, Hong Kong.

Larbi, E. (2012) *Public-Private Partnerships and the Poor in Water Supply Projects: The Ghanaian Experience*, WELL Fact sheet - Regional Annex, Ghana, www.lboro.ac.uk/well/resources/factsheets/factsheetshtm/RSA%20PPP%20and%20the%20oor.htm (accessed on 20 November 2012).

Laube, W. and van de Giesen, N. (2006) 'Ghanaian water reforms, institutional and hydrological perspectives', in Wallace, J. and Wouters, P. (eds) *Hydrology and Water Law – Bridging the Gap*, IWA Publishing.

Lazarte, E. (2011) *Private Small Water Supply Systems*, Handshake, International Financial Corporation (IFC), Washington DC.

Legislative Council (2008) *Hong Kong–Zhuhai–Macao Bridge, Hong Kong Boundary Crossing Facilities and the Link Road in Hong Kong*, LC Paper No. CB(1)1317/07–08(04), 25 April 2008, Legislative Council Panel on Transport, Hong Kong Special Administrative Region Government, http://www.thb.gov.hk/eng/whatsnew/transport/2008/200804252.pdf (accessed on 27 August 2008).

Leiringer, R. (2006) 'Technological innovation in PPPs: incentives, opportunities and actions', *Construction Management and Economics*, 24, 301–8.

Leisure and Cultural Services Department (2012) *Historic Building Appraisal*, http://www.lcsd.gov.hk/ce/Museum/Monument/form/Brief_Information_on_proposed_Grade_II_Items.pdf (accessed on 14 February 2012).

Leong, A. (2008) 'A sustainable West Kowloon cultural district', *Ming Poa Daily News*, 2 June 2008.

Levy, S. M. (1996) *Build Operate Transfer*, John Wiley and Sons, New York.

Li, B. (2003) 'Risk management of construction public private partnership projects', PhD thesis, Glasgow Caledonian University, UK.

Li, B., Akintoye, A., Edwards, P. J. and Hardcastle, C. (2004) 'Risk treatment preferences for PPP/PFI construction projects in the UK', *Proceedings: ARCOM Conference*, Heriot-Watt University, 1–3 September 2004, 2, 1259–68.

Li, B., Akintoye, A., Edwards, P. J. and Hardcastle, C. (2005a) 'The allocation of risk in PPP/PFI construction projects in the UK', *International Journal of Project Management*, 23(1), 25–35.

Li, B., Akintoye, A., Edwards, P. J. and Hardcastle, C. (2005b) 'Perceptions of positive and negative factors influencing the attractiveness of PPP/PFI procurement for construction projects in the UK', *Engineering, Construction and Architectural Management*, 12(2), 125–48.

Li, B., Akintoye, A., Edwards, P. J. and Hardcastle, C. (2005c) 'Critical success factors for PPP/PFI projects in the UK construction industry', *Construction Management and Economics*, 23, 459–71.

Li, H. and Liu, K. J. (2009) 'Research on risk evaluation of PFI projects with many-sided viewpoint', *Proceedings of CRIOCM 2009 International Symposium on Advancement of Construction Management and Real Estate*, 29–31 October 2009, Nanjing, China, 3, 1343–50.

Li, J., and Zou, P. X. W. (2011) 'A fuzzy AHP based risk assessment methodology for PPP projects', *Journal of Construction Engineering and Management*, ASCE, http://ascelibrary. org/coo/resource/3/jcemxx/256 (18 March 2011).

Liu, Y. W., Zhao, G. F. and Wang, S. Q. (2009) 'Case study VI – the National Stadium BOT project for Beijing 2008 Olympic Games', in Alfen, H. W., Kalidindi, S. N., Ogunlana, S. and Wang, S. Q. *Public Private Partnership in Infrastructure Development: Case Studies From Asia and Europe, EU–Asia Network of Competence Enhancement on Public Private Partnerships in Infrastructure Development*, Publisher of Bauhaus-Universität Weimar, Germany.

Liu, Z. Y. and Yamamoto, H. (2009) 'Public private partnerships (PPPs) in China: present conditions, trends and future challenges', *Interdisciplinary Information Sciences*, 15(2), 223–30.

Mak, C. W. (2008) 'The Chinese central government invests RMB7 billion in the Hong Kong-Zhuhai-Macau Bridge to lower toll fees', *Apple Daily* (Hong Kong), 6 August (in Chinese).

Mak, C. K., and Mo, S. (2005) 'Some aspects of the PPP approach to transport infrastructure development in Hong Kong', *Proceedings of the Conference on Public Private Partnerships – Opportunities and Challenges*, Hong Kong, 22 February 2005.

Maku, A. (2011) *Community-Public-Private Partnerships (CPPP) in the Three-District Water Supply Scheme*, TPP Project Fact Sheet, Ghana, www.ghana.watsan.net/page/777 (accessed on 20 October 2011).

Maltby, P. (2003) 'Has the PFI grown up?' *Public Finance*, London, August 2003.

Marin, P. (2009) 'Public-private partnerships for urban water utilities: a review of experiences in developing countries', *Trends and Policy Options*, No. 8, World Bank, Washington DC.

Merna, T. and Owen, N. (1998) *Understanding the Private Finance Initiative: The New Dynamics of Project Finance*, Asia Law and Practice Publishing Ltd, Hong Kong.

Ming Pao (2008a) 融資方案突變中央出資22億 (Chinese version only) English translation 'Sudden change in financing model, Chinese Central Government pays RMB2.2 billion', *Ming Pao Newspaper*, 6 August 2008, Hong Kong.

Ming Pao (2008b) 中央出資 港珠澳橋後年上馬港出67億 收費料可減至200元下 (Chinese version only) English translation 'Chinese Central Government invests in Hong Kong–Zhuhai–Macau Bridge, project commences in two years, Hong Kong invests RMB6.7 billion', *Ming Pao Newspaper*, 6 August 2008, Hong Kong.

Ming Pao (2010) 美荷樓申2億活化, (Chinese version only) English translation 'Mei Ho House revitalisation requests 0.2 billion', *Ming Pao Newspaper*, 21 April 2010, available at: http://hk.news.yahoo.com/article/100420/4/hlkz.html (accessed on 18 May 2010).

Ministry of Water Resources, Works and Housing (2011) *Water and Sanitation Sector Performance Report*, http://wsmp.org/downloads/4d8ca15ec1a12.pdf (accessed on 14 November 2011).

Mok, C. S. (2005) 'Privately financed infrastructure projects', MSc thesis, University of Hong Kong, Hong Kong.

Morgan, D. L. (1997) *Focus Groups as Qualitative Research*, Sage Publications, US.

Mustafa, A. (1999) 'Public-private partnership: an alternative institutional model for implementing the private finance initiative in the provision of transport infrastructure', *Journal of Project Finance*, 5(2), 64–79.

National Audit Office (2001) *Managing the Relationship to Secure a Successful Partnership in PFI Projects*, HC375, National Audit Office, UK, 29 November 2001.

National Audit Office (2011) *Lessons from PFI and Other Projects*, Report by the Comptroller and Auditor General, HC 920 Session 2010-2012, 28 April 2011.

National Treasury PPP Unit of South Africa (2007) *Public Private Partnership Manual*, http://www.treasury.gov.za/organisation/ppp/default.htm (accessed on 15 June 2007)

National Treasury of Republic of South Africa (2012) *About the PPP Unit*, http://www.ppp.gov.za/Pages/About.aspx (accessed on 20 November 2012).

New South Wales Government (2006) *Working with Government – Guidelines for Privately Financed Projects*, December 2006.

Ng, A. and Loosemore, M. (2007) 'Risk allocation in the private provision of public infrastructure', *International Journal of Project Management*, 25(1), 66–76.

Ng, S. T. and Wong, Y. M. W. (2006) 'Adopting non-privately funded public-private partnerships in maintenance projects: a case study in Hong Kong', *Engineering, Construction and Architectural Management*, 13(2), 186–200.

Nii Consult (2003) *Study on Provision of Services provided by GWCL – Final Report*, Accra.

Nijkamp, P., Van der Burch, M. and Vindigni, G. (2002) 'A comparative institutional evaluation of public-private partnerships in Dutch urban land-use and revitalization projects', *Urban Studies*, 39(10), 1865–80.

Nisar, T. M. (2007) 'Value for money drivers in public private partnership schemes', *International Journal of Public Sector Management*, 20(2) 147–56.

Nishimatsu (2006) *Nishimatsu and PFI*, Nishimatsu Construction Company Ltd, Hong Kong Branch.

Norusis, M. (2008) *SPSS 17.0 Guide to Data Analysis*, Pearson.

Nyarko, K. B. (2007) 'Drinking water sector in Ghana: drivers for performance', PhD thesis, UNESCO-IHE Institute for Water Education, Delft.

Obermauer, A. (2012) 'National railway reform in Japan and the EU: evaluation of institutional changes', *Japan Railway and Transport Review*, 29, December 2001, http://www.jrtr.net/jrtr29/pdf/f24_obe.pdf (accessed on 27 November 2012).

Ofosu, P. (2004) *Tariff and Water Cost: What Degree of Adequacy?*, The Union of African Water Supplies Congress, Accra.

Oriental Daily (2010a) 深水埗保育建築：北九龍裁判法院、美荷樓, (Chinese version only) English translation 'Redevelopment of Sham Shui Po: North Kowloon Magistracy,

Mei Ho House', *Oriental Daily*, 17 April 2010, http://orientaldaily.on.cc/cnt/finance/20100417/00205_004.html?pubdate=20100417 (accessed on 18 May 2010).

Oriental Daily (2010b) 兩億活化美荷樓, (Chinese version only) English translation '0.2 billion to revitalise Mei Ho House', *Oriental Daily*, 21 April 2010, http://the-sun.on.cc/cnt/news/20100421/00410_003.html?pubdate=20100421 (accessed on 17 May 2010).

Oriental Newspaper (2008) 花開堪折不去折直待無花空折枝 (Chinese version only) English translation 'Opportunity taken too late', *Oriental Newspaper*, 6 August 2008, Hong Kong.

Pantelias, A. and Zhang, Z. (2010) 'Methodological framework for evaluation of financial viability of public-private partnerships: investment risk approach', *Journal of Infrastructure Systems*, ASCE, 16(4), 241–50.

Partnerships UK (2012a) *Background*, http://www.partnershipsuk.org.uk/PUK-Background.aspx (accessed on 27 December 2012).

Partnerships UK (2012b) *What We Do*, http://www.partnershipsuk.org.uk/What-PUK-Do.aspx (accessed on 27 December 2012).

Partnerships Victoria (2000) *Partnerships Victoria Policy*, Department of Treasury and Finance, Victoria State Government, Australia.

Partnerships Victoria (2001) *Practitioner's Guide*, June 2001, Department of Treasury and Finance, Victoria State Government, Australia.

Partnerships Victoria (2008a) *Projects*, Department of Treasury and Finance, Victoria State Government, Australia, http://www.partnerships.vic.gov.au/CA25708500035EB6/WebProjects?OpenView (accessed on 24 July 2008).

Partnerships Victoria (2008b) *Policy and Guidelines*, Department of Treasury and Finance, Victoria State Government, Australia, http://www.partnerships.vic.gov.au/CA25708500035EB6/0/C0005AB6099597C2CA2570F50006F3AA?OpenDocument (accessed on 24 July 2008).

Partnerships Victoria (2008c) *Standard Commercial Principles*, April 2008, Department of Treasury and Finance, Victoria State Government, Australia.

PPP Canada (2012a) *What does the Canadian P3 Market Look Like?*, http://www.p3canada.ca/p3-market.php (accessed on 21 November 2012).

PPP Canada (2012b) *Overview*, http://www.p3canada.ca/about-ppp-canada-overview.php (accessed on 21 November 2012).

Pretorius, F., Lejot, P., McInnis, A., Arner, D. and Hsu, B. F. C. (2008) *Project Finance for Construction and Infrastructure – Principles and Case Studies*, Blackwell Publishing, Oxford.

Pribadi, K. S. and Pangeran, M. H. (2007) 'Important risks on public private partnership scheme in water supply investment in Indonesia', *Proceedings of the 1st International Conference of European Asian Civil Engineering Forum*, EACEF 31, Indonesia, 26–27 September 2007.

Price, W. (2002) 'Innovation in public finance', *Public Works Management and Policy*, 7(1), 63–78.

Project Finance (2007) 'Skies not the limit', *Project Finance*, April 2007.

Qi, X., Ke, Y. J. and Wang, S. Q. (2009) 基于案例的中国PPP项目的主要风险因素分析 (Chinese version only) English translation 'Analysis of critical risk factors causing the failures of China's PPP projects', *China Soft Science*, 221, 107–13.

Qiao, L., Wang, S. Q., Tiong, R. L. K. and Chan, T. S. (2001) 'Framework for critical success factors of BOT projects in China', *Journal of Project Finance*, 7(1), 53–61.

Qiu, Q. (2008) 'Pact inked of funding Pearl River Delta Bridge', *China Daily Hong Kong Edition*, 6 August 2008, People's Republic of China.

Qu, Y. and Li, J. (2009) 'The research on PPP model for low-rent housing of China', *Proceedings of CRIOCM 2009 International Symposium on Advancement of Construction Management and Real Estate*, 29–31 October 2009, Nanjing, China, 2, 591–6.

Queensland Government (2008a) *Business Case Development*, http://www.dip.qld.gov.au/docs/library/pdf/ppp/ppp_guide_bus_case_dev.pdf (accessed on 29 July 2008).

Queensland Government (2008b) *Southbank EPI Centre*, http://www.southbank.tafe.net/site/epicentre/ (accessed on 28 March 2008).

Regan, M., Smith, J. and Love, P. (2009) 'Public private partnerships: what does the future hold?', *Proceedings of the RICS COBRA Research Conference*, Cape Town, 10–11 September 2009, 462–74.

Roads Traffic Authority (2003) *Cross City Tunnel: Summary of Contracts*, Roads and Traffic Authority of New South Wales Government, Australia.

Roads Traffic Authority (2007) http://www.rta.nsw.gov.au/constructionmaintenance/major-constructionprojectssydney/crosscitytunnel/index.html (accessed on 30 May 2007).

Roger, B. (2009) *PPPs and the Water Sector: Plugging the Infrastructure Hole*, Deloitte, Australia.

Sachs, T., Tiong, R. and Wang, S. Q. (2007) 'Analysis of political risks and opportunities in public private partnerships (PPP) in China and selected Asian countries', *Chinese Management Studies*, 1(2), 126–48.

Sapte, W. (1997) *Project Finance: The Guide to Financing Build Operate Transfer Projects – Uses in PPP*, Euromoney, Hong Kong.

Satpathy, I. and Das, B. (2007) 'Sustainable strategy and policy making module on infrastructure development via PPP mechanisms: a perspective for application in India', *Proceedings of the 2007 International Conference on Concession Public/Infrastructural Projects (ICCPIP)*, Dalian University of Technology, Dalian, China, 24–26 August 2007.

Savannah College of Art and Design (2012a) *About SCAD – History*, http://www.scad.edu/about/history.cfm (16 February 2012).

Savannah College of Art and Design (2012b) *Savannah*, http://www.scad.edu/savannah/index.cfm (16 February 2012).

Savannah College of Art and Design (2012c) *About SCAD – Facts*, http://www.scad.edu/about/facts.cfm (16 February 2012).

Sharma, S. (2007) 'Exploring best practices in public–private partnership (PPP) in e-Government through select Asian case studies', *International Information and Library Review*, 39(3–4), 203–10.

Shen, L. Y. and Wu, Y. Z. (2005) 'Risk concession model for BOT contract projects', *Journal of Construction Engineering and Management*, ASCE, 131(2), 211–20.

Shen, L. Y., Platten, A. and Deng, X. P. (2006) 'Role of public private partnerships to manage risks in public sector projects in Hong Kong', *International Journal of Project Management*, 24(7), 587–94.

Sing Tao Daily (2009), 八和要「申冤」披露方案作公論 (Chinese version only) English translation 'Chinese artists association should let public judge its proposal', *Sing Tao Daily*, 23 February 2009.

Smith, H. (2012) *Public Private Partnerships (PPPs) in Hong Kong*, http://www.cannonway.com/web/print.php?page=96 (accessed on 11 December 2012).

So, U. (2009) 'West Kowloon hub empowered to cover costs', *The Standard*, 11 September 2007, http://www.thestandard.com.hk/news_print.asp?art_id=53223&sid=15309731 (accessed on 17 March 2009).

South China Morning Post (2008) 'Now we need action on the arts hub project', 19 June 2008.

Sun, Y., Fang, D. P., Wang, S. Q., Dai, M. D. and Lv, X. Q. (2008) 'Safety risk identification and assessment for Beijing Olympic Venues Construction', *Journal of Management in Engineering*, 24(1), 40–7.

Tai Kung Pao (2010a) 中區警署新設計七月公布 (Chinese version only) English translation 'New design of Victoria Prison released in July', *Tai Kung Pao*, 28 April 2010, available at: http://www.takungpao.com.hk/news/10/04/28/GW-1249819.htm (accessed on 18 May 2010).

Tai Kung Pao (2010b) 雷生春賣涼茶每碗僅售五元 – 現金券作招徠 星展稱無誤導 (Chinese version only) English translation 'Lei Sun Chung sells bitter tea at five dollars per bowl – cash coupon reward, DBS bank confirms', *Tai Kung Pao*, 21 April 2010, available at: http://www.takungpao.com.hk/news/10/04/21/images_0708-1246631.htm (accessed on 19 May 2010).

Tam, C.M. (1999) 'Build-operate-transfer model for infrastructure developments in Asia: reasons for successes and failures', *International Journal of Project Management*, 17(6), 377–82.

Tam, C. M., Li, W. Y. and Chan, A. P. C. (1994) 'BOT applications in the power industry of South East Asia: a case study in China', *East Meets West*, Procurement Systems Symposium CIB W92 Proceedings Publication, 175, 315–22.

Thomas, A. V., Kalidindi, S. N. and Ananthanarayanan, K. (2003) 'Risk perception analysis of BOT road project participants in India', *Construction Management and Economics*, 21(4): 393–407.

Thomas, A. V., Kalidindi, S. N. and Ganesh, L. S. (2006) 'Modeling and assessment of critical risks in BOT road projects', *Construction Management and Economics*, 24(4): 407–24.

Tieman, R. (2003) 'A revolution in public procurement: UK's private finance initiative', *Financial Times*, London, 24 November 2003.

Tiong, R. L. K. (1996) 'CSFs in competitive tendering and negotiation model for BOT projects', *Journal of Construction Engineering Management*, 122(3), 205–11.

Townsend, I. (2004) 'Springborg envisions united conservative party', AM ABC Radio, 13 November 2004, http://www.abc.net.au/am/content/2004/s1242839.htm (accessed on 28 March 2008).

Trafford, S. and Proctor, T. (2006) 'Successful joint venture partnerships: public-private partnerships', *International Journal of Public Sector Management*, 19(2), 117–29.

Training Research and Networking for Development (2003) *Water, Sanitation and Service Delivery in Ghana*, WELL Fact Sheet, www.trend.watsan.net (accessed on 21 January 2012).

Transport and Housing Bureau (2008a) *Hong Kong–Zhuhai–Macao Bridge*, Hong Kong Special Administrative Region Government, 7 March 2008, http://www.thb.gov.hk/eng/policy/transport/issues/cbt_3.htm (accessed 12 August 2008).

Transport and Housing Bureau (2008b) *Consensus reached on financing of HK-Zhuhai-Macao Bridge*, Hong Kong Special Administrative Region Government, 28 February 2008, http://www.thb.gov.hk/eng/psp/pressreleases/transport/land/2008/200802291.htm (accessed on 12 August 2008).

Transport and Housing Bureau (2008c) 運輸及房屋局局長談港珠澳大橋 (答問部分) (Chinese version only) English translation 'Secretary of Transport and Housing Bureau speaks of the Hong Kong–Zhuhai–Macau bridge (questions and answers)', Hong Kong Special Administrative Region Government, 28 February 2008, http://www.thb.gov.hk/tc/psp/pressreleases/transport/land/2008/200802293.htm (accessed on 12 August 2008).

Tuffour, B. (2011) 'Public-private partnerships model in small towns O&M contract in Bekwai, Atebubu and Wassa Akropong', www.ghana.watsan.net/page/777 (accessed on 20 November, 2011).

Udemezue, O. C. (2012) 'Nigeria: rescuing development through PPP techniques (1) Business day', http://www.businessdayonline.com/NG/index.php/analysis/commentary/36601-nigeria-rescuing-development-through-ppp-techniques-1 (accessed on 20 November 2012).

United Nations Economic Commission for Europe (2004) *Governance in Public Private Partnerships for Infrastructure Development*, UNECE, Geneva.

Unkovski, I. and Pienaar, E. (2009) 'Public private partnerships in South Africa: analysis and management of risks', *Proceedings of the Construction and Building Research Conference of the Royal Institute of Chartered Surveyors* (CORBRA 2009), University of Cape Town, 10–11 September 2009.

Van der Kamp, J. (2008) 'New funding proposal made for Zhuhai Bridge', *South China Morning Post*, Hong Kong, 1 August 2008.

Van-Ess, R. K. D. (2009) 'Ghana – role of private operator in small towns water supply', World Bank Water Week, Community Water and Sanitation Agency (CWSA), 17–20 February, 2009, Ghana.

Vaughn, S., Schumm, J. S. and Sinagub, J. (1996) *Focus Group Interviews in Education and Psychology*, Sage Publications, US.

Walker, C. and Smith, A. (1995) *Privatized Infrastructure: The BOT Approach*, Thomas Telford, London.

Wang, S. Q. and Ke, Y. J. (2009) 'Case study IV – Laibin B power project – the first state-approved BOT project in China' in Alfen, H. W. (ed.) *Public-Private Partnership in Infrastructure Development* (first edition), Bauhaus-Universität Weimar.

Water and Sanitation Program (2010) *Public-Private Partnerships for Small Piped Water Schemes – A review*. World Bank, Washington DC.

WaterAid (2010) *Urban Water Sector Assessment*, www.wateraid.org (accessed on 2 October 2010).

WaterAid (2011) *Urban Water Sector Assessment*, www.wateraid.org (accessed on 10 October 2011).

Wong, A. (2007) 'Lessons learned from implementing Infrastructure PPPs – a view from Singapore', Seminar jointly organised by the Department of Civil Engineering of the University of Hong Kong and Civil Division of the Hong Kong Institution of Engineers, 13 June 2007.

Wong, A. (2008) 'ICAC to take role in public private deals', *The Standard*, 31 October 2005, http://www.thestandard.com.hk/news_print.asp?art_id=4529&sid=5255280 (accessed on 20 August 2008).

World Bank (2011a) *World Development Indicators*, www.google.com.gh/publicdata (accessed on 12 November 2011).

World Bank (2011b) 'Management contract for urban water between Ghana Water Company Ltd and –X', Draft 1, December 2004, http://siteresources.worldbank.org/INFOSHOP1/Resources/ManagementContractGhana.pdf (accessed on 14 November 2011).

World Bank (2012) Private Participation in Infrastructure Database, http://ppi.worldbank.org/explore/ppi_exploreCountry.aspx?countryID=50 (accessed on 15 January 2012).

Wibowo, A. and Kochendörfer, B. (2005) 'Financial risk analysis of project finance in Indonesian toll roads', *Journal of Construction Engineering and Management*, 131(9): 963–72.

Wikipedia (2013) File: Western Harbour Tunnel.JPG, http://en.wikipedia.org/wiki/File:Western_Harbour_Tunnel.JPG#filelinks (accessed on 28 January 2013).

Wu, E. (2008) 'HK$21.6b approved to bankroll arts hub', *South China Morning Post*, 5 July 2008.

Xenidis, Y. and Angelides, D. (2005) 'The financial risks in build-operate-transfer projects', *Construction Management and Economics*, 23(3), 431–41.

Xu, Y. L., Yeung, J. F. Y., Chan, A. P. C., Chan, D. W. M., Wang, S. Q. and Ke, Y. J. (2010) 'Developing a risk assessment model for PPP projects in China: a fuzzy synthetic evaluation approach', *Automation in Construction*, 19(7), 929–43.

Yahoo! Finance (2012) Currency Converter, http://hk.finance.yahoo.com/currencies/conv erter/#from=HKD;to=HKD;amt=1 (accessed on 11 December 2012)

Yescombe, E. R. (2008) *Public Private Partnerships – Principles of Policy and Finance*, Elsevier, UK.

Zaato, J. J. (2011) 'Contractualism as a reform and governance tool: a critical analysis of urban water management reforms in Ghana', The Canadian Political Science Association Annual Conference, Wilfred Laurier University, 16–18 May 2011.

Zhang, X. and AbouRisk, S. S. (2006) 'Relational concession in infrastructure development through public-private partnerships', *Proceedings of the CIB W89 International Conference on Building Education and Research*, CIB, Hong Kong, 10–13 April 2006.

Zhang, X. Q. (2001) *Procurement of Privately Financed Infrastructure Projects*, PhD thesis, University of Hong Kong, Hong Kong Special Administrative Region.

Zhang, X. Q. (2005a) 'Critical success factors for public–private partnerships in infrastructure development', *Journal of Construction Engineering and Management*, 131(1), 3–14.

Zhang, X. Q. (2005b) 'Paving the way for public–private partnerships in infrastructure development', *Journal of Construction Engineering and Management*, ASCE, 131(1), 71–80.

Zhang, X. Q. (2006) 'Factor analysis of public clients' best-value objective in public–privately partnered infrastructure projects', *Journal of Construction Engineering and Management*, ASCE, 132(9), 956–65.

Zhong, L. and Fu, T. (2010) *BOT Applied in Chinese Wastewater Sector*, http://www.adb.org/ Water/actions/prc/PRC-BOT-wastewater-paper.pdf (7 October 2010).

Zhong, L., Mol, A. P. J. and Fu, T. (2008) 'Public private partnerships in China's urban water sector', *Environmental Management*, 41, 863–77.

Index